THE
G.O.D.
EXPERIMENTS

THE
G.O.D.
EXPERIMENTS

———◆———

*How Science Is Discovering God
in Everything, Including Us*

Gary E. Schwartz, Ph.D.

with William L. Simon

ATRIA BOOKS

NEW YORK   LONDON   TORONTO   SYDNEY

ATRIA BOOKS

1230 Avenue of the Americas
New York, NY 10020

First **ATRIA** BOOKS trade paperback edition May 2007

ATRIA BOOKS and colophon are registered trademarks of Simon & Schuster, Inc.

For information about special discounts for bulk purchases,
please contact Simon & Schuster Special Sales at
1-800-456-6798 or business@simonandschuster.com.

Manufactured in the United States of America

10 9 8 7 6 5 4 3 2 1

The Library of Congress has cataloged the hardcover edition as follows:

Schwartz, Gary E.
The G.O.D. experiments : how science is discovering God in everything, including us /
Gary E. Schwartz with William L. Simon—1st Atria Books hardcover ed.
p. cm.
Includes bibliographical references (p.   ) and index.
1. Intelligent Design (Teleology) I. Title: G.O.D. experiments. II. Simon, William L.
III. Title.

BL262.S37 2006
212'.1—dc22

ISBN-13: 978-0-7434-7740-6
ISBN-10:   0-7434-7740-5
ISBN-13: 978-0-7434-7741-3 (pbk)
ISBN-10:   0-7434-7741-3 (pbk)

To our parents
and to Susy, Willie, and Sam
Arynne, Victoria, and Sheldon
and Vincent and Elena

The important thing is not to stop questioning.

ALBERT EINSTEIN

CONTENTS

Can Science Take Us to God?

Imagine that there had been no Abraham, no Moses, no Jesus, no Muhammad. Imagine a world without mystics, prophets, shamans, or medicine men. And what if there were no Old Testament, no New Testament, no Koran—just contemporary, twenty-first-century science? In such a world, would scientists feel the need to come up with a theory that some kind of invisible organizing process—that is, a God—exists in the universe and in our daily lives? A rumor would leak to the media, stirring a high level of skeptical coverage, on the offbeat notion that science had proposed the existence of a universal intelligence, a source of cosmic energy, that explained the origin and evolution of order in the universe.

Let's face it: not in that fictional world but in today's real world, where some version of the God concept is accepted by a majority of the planet's 6 billion people, God is a subject traditionally shunned by scientists. And for good reason: a sure way for federally funded scientists to lose their government support from the National Institutes of Health, the National Science Foundation, or NASA is for them to start conducting experiments documenting the existence of God.

But is the belief in God—a universal intelligence, source and energy of all that is—something we must accept only on faith? Or is there compelling scientific reasoning, supported by incontrovertible experimental evidence? If this essence is actually at work in the scientist's lab as well as in our daily lives, it's time for us to take notice. My experiences in the laboratory and in life demonstrate convincingly, I believe, that science can lead us to the God who is now making himself/herself/itself known in physics, statistics, computer science, and even in, of all places, parapsychology experiments. We no longer need to accept God on faith alone. This news won't rock the boat of people for whom faith has been and always will be sufficient. But even the committed will, I think, find the support for their beliefs persuasive. And for everyone else, the results of the experiments and research presented in these pages open a world of new possibilities and new understanding.

GOD AND G.O.D.

The everyday word "God," and associated terms like "Lord," "King of the Universe," "the Ultimate," "the Source," "the Great Spirit," and "Allah," have different meanings for different religions as well as for different people within a given religion.

For some, God is a demanding older white male who sports a Santa-like beard and leans upon a gnarled and weathered staff. For others, a nurturing female deity who plays a sitarlike instrument . . . or a spirit that animates everything from elementary particles to galaxies . . . or a cosmic consciousness that exists beyond space and time. These widely diverse images of God—from the highly specific to the extraordinarily broad—have through the ages created confusion and antagonism, suspicion, hatred, and bloodshed. Little wonder that most scientists avoid discussing God in the context of their work. Little wonder that any social discussion of God, except among fellow-believers, is considered as unacceptable as bragging about the accomplishments of one's favorite sports team.

To avoid all the emotional baggage that too often comes into play when somebody tries to tell you about his or her notion of God, in these pages you'll find, instead, the term "G.O.D.," which I use to suggest a Guiding, Organizing, Designing process. Where you find "G.O.D" or "the G.O.D. process," you can decide to accept it as referring to your own idea of God, or as specifically representing a *Guiding-Organizing-Designing* process (the precise meaning in this book). I use the term "G.O.D." as a specific concept that everyone can focus on and discuss—even scientists.

How the concept of G.O.D. relates to your personal idea of "God" will become clear to you as you read this book. In a confusing and troubling world, now more than ever a meaningful question for all of us to consider is whether a G.O.D. process plays an invisible role in the evolution of our planet, in the unfolding of our universe, and in our personal lives. This is a central question you'll find addressed in these pages—the question of intelligent evolution.

I BECOME INVOLVED

I was educated to be an open-minded skeptic. As a Harvard Ph.D. student I was taught how to raise questions about mind and health, convert them into stable hypotheses, design experiments to critically evaluate these hypotheses, and then follow the findings where they took me. I was not trained to consider the nature of God from a scientific point of view.

For twenty years I was a mainstream academic psychophysiologist who conducted research and applications in the field of mind-body medicine. I am a scientist—originally on the faculty at Harvard, later on the faculty at Yale as a tenured professor at the age of thirty-two. My work has been funded by the National Science Foundation and the National Institutes of Health, and I'm currently a professor of psychology, surgery, medicine, neurology, and psychiatry at the University of Arizona.

My training never taught me to ask questions that would lead me to think about God, and I became involved with the G.O.D. hypothesis only because my research simply led me there. Ultimately I couldn't be true to my professional responsibilities as a scientist and educator—or be true to myself—if I did not publish what I had discovered in my research and laboratory work about G.O.D.'s existence in our daily lives.

Though Darwinian evolutionists may wish to paint this book as serving the Christian right, and creationists may choose to embrace it as supporting their literalist theist views, the truth is that this book was written by an agnostically raised scientist with an open and inquisitive mind who is following the data where it takes him.

CAN SCIENCE TAKE US TO G.O.D?

Imagine that science can actually take us to G.O.D.—not merely theoretically but experimentally. Imagine that science can help us discover how a G.O.D. process operates not only in the creation and evolution of molecules, organisms, planets, stars, and galaxies, but in the manifestation of our most significant personal relationships. Imagine that this science-based G.O.D. is with you as you read this book, playing a behind-the-scenes role in your most intimate thoughts and feelings. *Imagine that this G.O.D. is literally inside you right now, patiently waiting for you to wake up to this inner, ultimate reality.* Are you ready to wake up?

Part One

THE SCIENCE OF
PROPHECY

What experiments with a British psychic who claims to see the future in his dreams reveal about the existence of G.O.D.

Extraordinary claims require extraordinary evidence.

CARL SAGAN

1

Foreseeing God in the Laboratory

HOW EVIDENCE FOR DESIGN SHOWS UP IN PROPHETIC DREAMS

One April day in 2001, the phone in my home office rang. The caller was a man with a charming British accent and a sparkling manner alive with animation and humor, who introduced himself as Chris Robinson. He had read about my lab at the University of Arizona, he said, and was calling to announce that he wanted me to test his abilities to see if I could conclude that he was, indeed, getting tips from some otherworldly source.

After a near-death experience thirteen years earlier, weird messages had begun arriving in his sleep—dreams that foretold the future, especially about murders and terrorism.

Over the years, he said, evidence obtained through his dreams had helped put many criminals behind bars. Because of Robinson's information, murderers who thought they had escaped were caught, IRA bombers were captured, and corrupt members of the police force were uncovered and sent to prison. Fingering corrupt cops and detectives didn't make Christopher popular with British law enforcement, he said. Although they had continued to listen to his information and act on it, he said, he had been kept at arm's length and mistrusted.

I had no reason to think anyone could really do what Christopher was claiming. I listened patiently as he shared stories that were amazing and outrageous. I found most of what he said virtually impossible to believe. He claimed that a book called *Dream Detective* had been published in England about many of his cases and that he would send me a copy. The book actually arrived, and I found it supported his claims, apparently substantiating his uncanny skill.

I was about to start on a journey on which this total stranger from England would ultimately propel me to reconsider the entire history of my scientific career, and in the process come to new and meaningful conclusions.

We had a series of phone calls after that. Trained in clinical psychology, I listened closely for signs of psychosis or thought disorder. He not only sounded sane but was insisting that he fly over from England if I would conduct tests to verify his claims of dreams that foretold the future. This blue-collar worker with marginal income was offering to buy his own airline ticket and pay for his own hotel and meals if I was willing to try to help him find out, once and for all, what his power really represented.

Over the next weeks Christopher and I discussed a highly controlled yet seemingly impossible (to me, not to Christopher) experiment that begged to be conducted. I realized that if Christopher was neither a delusional schizophrenic nor a pathological liar—these were two big ifs—and the findings were positive, the experiment held the possibility of becoming one of the more remarkable investigations in the history of contemporary parapsychology, and perhaps even of science in general. I subsequently learned that Christopher could be unreasonably suspicious at times—no doubt because of his dangerous work as an undercover agent and his extraordinary sensitivity as a psychic.

Four months later, in early August 2001, Christopher arrived from

England and set up temporary residence in a Tucson hotel. I had by then selected twenty possible locations in southern Arizona—from Nogales, a Mexican border town, to Summerhaven, a ski resort on the top of Mount Lemmon. Of these twenty locations known only to me, ten would be selected at random for us to visit on ten successive days.

I printed out the name of each site on a sheet of paper, placed each sheet in a separate envelope, sealed the envelopes, and shuffled them, then shipped them overnight to my friend and coauthor Bill Simon, who had agreed to help in the experiment. He acted as an intermediary, receiving the package and turning it over to a third party whose identity was unknown to me.

The third party was instructed to open the package, shuffle the envelopes, number them, and store them in a safe place; all of this was to be done in front of a video camera.

For thirteen years Chris Robinson had been recording his nightly dreams and premonitions in a diary—his "dream diary." Since the previous May, anticipating the experiment that he and I had been discussing, Chris recorded his dreams for ten nights about what he thought he would do and see on each day in Tucson. He repeated this ten-night pre-experimental dream sequence in June and again in July. When he arrived in Tucson in early August, he brought the three sets of ten-day dream diaries with him. I read the diary with its predictions, still believing this part of the experiment to be preposterous and fully expecting that nothing would come of it. Even by my adventurous standards, it simply seemed impossible that he could have guessed locations anything like the ones we would visit on each day of the experiment.

THE "TEN DAYS IN ARIZONA" EXPERIMENT

In his hotel across the street from the University of Arizona, Christopher got ready for bed on the night before the first day of our experiment. Before falling asleep, he would ask the universe, in his head, to be shown in his dreams where he would be taken the next day.

As was his ritual, when he awoke, he wrote careful notes in his diary about the dreams he'd had, often including sketches or diagrams. On the first morning—Thursday, August 2, 2001—the experiment officially began when I arrived, about 9 A.M. As we said our good mornings, I set up my digital video camera and videotaped the pages Christopher had written, while he read the raw information out loud onto the audio track. On camera, he then summarized the key information from the three pre-experiment dreams for that day. After that, he described the previous night's dreams. And then, the key part, he concluded by summarizing from the dreams his description of things he believed we would see or experience during that one day.

On the first day, Robinson focused on "holes, lots of holes," along with "a basin empty of water." With the camera still running, I then placed a call to Bill Simon in southern California to tell him we were ready to learn the location for the day. Bill then contacted the third party. (He didn't have to go far; I would learn after the experiment that this mysterious third party was Bill's wife, Dr. Arynne Simon.) With her video camera running, she opened the envelope that was at the top of the stack after her shuffle, which she had marked as Envelope #1. She read out loud what was written on the paper: "Desert Museum/Animals." She showed the paper to the camera and to Bill, who then called me back and told me the location.

I did not tell Christopher where we would be going, nor did I tell Bill what Christopher had dreamed. But I knew the desert museum well: a place with a huge variety of holes, ranging from human caves and large animal cages in the ground to prairie dog tunnels in every direction. Even before Christopher and I packed our assorted video and still cameras and began to drive out of Tucson in the direction that would take us to the museum, I knew that his prediction of "holes, lots of holes" was a remarkably apt description of the landscape of the museum. Also, the museum was located in a basin that millions of years ago had been an ocean.

Christopher's dreams for each day included information not just about the site but also about objects and events on the journey to the location. Hence we carefully monitored the journeys as well as the sites.

For example, on Day 2, Christopher said that the primary themes of his dreams were "shops and workshops . . . fabricating things . . . metal." The secret message in the envelope for that day sent us to Tubac, an artists' colony, to a specific shop displaying metal sculptures, and with a workshop in the back. On Day 4, Christopher said that the primary themes of his dreams were "suns, mirrors, LCDs, telescopes, Mount Olympus [after his 35 mm camera], airplanes, hangars, a pitched propeller." The site for the day was Kitt Peak National Observatory, situated on a mountaintop and housing the world's largest solar telescope. Returning, we stopped for lunch at one of the only places available on that mountain road—a general aviation airport, where we of course saw hangars; as decoration, the airport restaurant featured prominently out front a large pitched propeller!

The post-location information was extra—not part of the main experiment. However, it turned out that the "extra" information was also extra in the sense of being truly "extra-ordinary." The late Carl Sagan was fond of saying, "Extraordinary claims require extraordinary evidence." Sagan's slogan, which has become one of the mantras of my Human Energy Systems Laboratory, occurred to me often during those ten days.

It turned out that there were degrees of extra-ordinariness in the findings from this ten-day experiment. The results every day proved extraordinary, but some days proved to be beyond extraordinary; I want to share with you two in particular—chosen partly because they are typical, partly because they are profound, and partly because some of the story is playful.

A REPRESENTATIVE DAY—FROM THE BORDER TO THE "SPIRIT OF GOD"

Day 5 started with Christopher listing the highlights from his previous night's dreams, informed by observations from his prior Day 5 dreams collected at his home in May, June, and July.

He had dreamed of men holding up traffic. This scene was set somewhere near a border, he said, taking for granted it was the Mexican border, not far from Tucson. He also saw a large water or gas tank.

He had dreamed of boats, many boats, and of a car with four flat tires. He had written in his dream journal that the car had no "oil" and then he added "mineral oil." He also saw an embassy in London.

He dreamed of tires piled high along a chain-link fence, and had drawn a picture of the tires and fence in his diary. He dreamed of my needing to use a credit card to get into a store or building. He also saw loads of umbrellas. And something about the "Spirit of God."

As I listened to this jumbled laundry list of items, I wondered: would the envelope, when it was opened, reveal that we were going to the Mexican town of Nogales—one of the twenty locations I had provided? It would fit many of the fundamentals in the dreams—men holding up traffic, border, water tanks, boats (people haul boats back and forth to Mexican beach towns), cars with flat tires, old tires along fences, umbrellas.

After all the information was recorded, I called to say I was ready to know the location. Bill called back a few minutes later to announce, "Gary, you're going to 'the old gem and mineral store.'"

From the first, I had not expected Christopher to get *anything* right beyond a few lucky guesses. The previous four days had been eye openers—his predictions had been remarkably on target. But it was too much to expect that winning streak to continue, and when I heard the location, it was clear that this fifth day was going to break his string of home runs. I reviewed in my mind the things Christopher had described, and couldn't come up with a single thing that we might see at the gem and mineral store, or that we might see on the way there or back.

Of course, I didn't tell Christopher what I was thinking. Moreover, following the experimental protocol, I did not tell Christopher where we were going.

Off we went. After about ten miles, as we passed an area of homes and a shopping mall surrounded by multiarmed cacti reaching twenty or more feet in the air, Christopher called my attention to some men on a side street, holding up traffic. He asked, "Are we near the Mexican border?"

"No," I told him. "We're more than forty miles from the border."

"Are you sure?" he asked. "My dream was very specific that men would be holding up cars near a border."

I said, "Christopher, trust me—the Mexican border is far from here." But to humor him, I maneuvered to see what was causing the tie-up. Workmen holding up traffic are commonplace enough; I simply assumed that this was not evidential. I was wrong.

We discovered there was construction work being done on the road. Beyond the work crew, I spotted a large water tank. And much to my amazement, the tank bore a painted sign that read "Borderland."

Borderland! We were not near an actual border, but we had encountered the word, in large, hard-to-miss lettering. Was I stretching the evidence to make the dreams fit the circumstances? Quite possibly, I reminded myself.

A bit farther on, a pickup truck went by us, pulling a boat on a trailer. Moments later we passed a boat storage yard. As I made a turn onto the street of the gem and mineral store, we saw eight more boats stored near the roadway. That's a lot of boats in the desert, I thought to myself.

We reached the gem and mineral store, and sitting on the right side of the parking lot, I am almost too embarrassed to admit, was a car with four flat tires! I couldn't believe it.

Christopher got all excited. He screamed, "Gary, this is just like how I dreamt it. In my dream I was able to walk right up to the tires and touch them." I took pictures of Christopher touching the flat tires.

Christopher had dreamed of some connection between the car and an embassy. The car with the flat tires turned out to be an old Rambler Ambassador. Embassy, ambassador. Another stretch? The car, obviously not driven for years, clearly had no oil—and obviously no "mineral" oil (though it was parked in front of a gem and mineral store).

But none of Christopher's other items connected. We did not see tires piled along a chain-link fence. I did not need a credit card to enter the store. There were no umbrellas at the location. And I did not see anything that suggested a "Spirit of God" connection.

We returned to Christopher's hotel and checked off what information in his dream was related to the journey and the location, and what was not. My expectation that the gem and mineral store would not connect with any of his dream clues was clearly wrong—he did much better than I expected. However, he obviously had not been perfect. The experiment was officially over for the day. But now is when the story gets even weirder. Yet it's all true—it really happened, and it's recorded on videotape.

FINDING THE "SPIRIT OF GOD" AT COSTCO?

I discovered that I was running out of videotapes for my mini DV camera, and Christopher was running out of them for his camera as well. I had one hour of free time before I was due at the laboratory, and suggested to Christopher that I take him to a store where we could purchase videotapes in bulk. Ever flexible, Christopher said, "Sure, let's go."

I took him to Costco, a huge discount store that typically sells products in large volume. To get in, I had to show my Costco membership card. Christopher tried to claim another hit: "Look, Gary, just like my dream predicted—you need a credit card to get into a store!" I thought to myself, This is probably because he already mentioned that I would be using a credit card for this purpose, and it unconsciously affected my choice of where we would come to buy the tapes. I dismissed the observation as mere suggestion.

As we were ready to leave, Christopher said, "Gary, I'm hungry. Can we get something to eat?" I explained that I had to get to an appointment at the lab, and we would have time only if we grabbed something quickly. Christopher said, "Sure, let's eat here." At the Costco fast food counter, I bought him a chicken and cheese sandwich, and I sat down with him to wait while he ate. And my world turned upside down. Christopher pointed off to his right and said, "Look, Gary . . ." There, clear as day, were piles of tires placed along a chain-link fence, just as he had drawn them in his dream journal.

I was wide-eyed. But there was more. He smiled and gestured all around us. Seemingly everywhere were large umbrellas. There we were, sitting inside a huge building around plastic outdoor tables with huge shade umbrellas, at least fifteen of them. I stared at these umbrellas in disbelief.

And then I noticed that there was writing on the umbrellas. A company had seen fit to decorate them with advertising. What I read on those umbrellas that day I will remember for the rest of my life. The advertising was the slogan for Hebrew National hot dogs: "We answer to a Higher Authority." And Christopher had predicted we would encounter something about the "Spirit of God."

Christopher's dreams had told him not only that we would be visiting a place that would have a car with four flat tires (an obviously highly specific and unique piece of information) and no oil, but that I would be taking us to a store requiring my (credit) card to enter. And that store had finally brought us face-to-face with "We answer to a Higher Authority."

Give me a break! My mind boggled in confusion and conflict; from the observations, I began to understand what spiritual people mean when they say, "There are no coincidences." How could Christopher know these things?

A skeptic once said to the distinguished anthropologist Margaret Mead, "These are the kind of data I wouldn't believe even if they were true!" I recalled this phrase as I wondered what else could possibly happen in the remaining five days of this experiment. What actually happened was beyond anything I could have predicted.

It's too coincidental to be accidental.

Susy Smith

2

Discovering Intelligent Design in Our Lives

A STUNNING CONCLUSION TO AN EXTRAORDINARY PSYCHIC EXPERIMENT

The question arises, was there convincing evidence for extraordinary "intelligent design" not only in conducting this unique precognitive experiment, but also in carefully arranging the seemingly random order of the locations to fit the detailed personal schedules of our complex lives? In other words, did a "higher authority" invisibly arrange the selection of the sites to fit Christopher's personal schedule as well as Bill Simon's and my professional schedules?

Was there a sophisticated yet discernable pattern to the arrangement of the entire experiment that somehow managed to coordinate our three respective schedules, even though the locations had been supposedly randomly arranged, first when I shuffled the envelopes, and again when Dr. Arynne Simon did?

JUMPING TO DAY 9—OUR SCHEDULE FITS TOO WELL

Bill Simon wanted to meet Christopher, to see with his own eyes how the experiment was being conducted. He wanted to make sure that there was no hanky-panky in the procedure—no deliberate or unintentional fraud being performed by Christopher or (even though he knows me and vows that he trusts my scientific integrity) by me. Since earlier in life he trained and worked briefly as a magician, he might see something I had been missing. Bill decided to fly to Tucson to witness Day 10 of the experiment firsthand.

When I went to San Diego to work on a book with him, Bill or his wife, Arynne, picked me up at the airport. He asked me if I would be able to do the same for him.

I explained that it depended upon which location was revealed in the envelope to be opened that day. If the location turned out to be close to the airport, I could probably meet his flight. But if the location was far from the airport, I wouldn't have time.

Only I knew what the twelve remaining locations were, and a number of them were too far away.

On the morning of Day 9, I was with Christopher in his hotel, videotaping the information for the day. Christopher saw parking meters in his dreams. He saw a large number of satellite dishes. He saw a woman being murdered later in the day, not far from where we were going. He also saw three people driving from an airport. He saw planes like "arrowheads" taking off and landing.

As I heard this information, I was aware that parking meters, satellite dishes, and a woman being murdered could fit the town of Nogales, one of the remaining locations. If so, then the three people he saw driving from the airport, and the arrowhead planes he saw landing (Tucson has a large Air Force base where small, dark, arrow-shaped fighter jets regularly take off and land), could not represent Christopher and me after we had picked up Bill at the Tucson airport: Nogales would be too far away.

When the Day 9 envelope was opened, I learned that Christopher and I were not going to downtown Nogales, we were going to downtown Tucson—in fact, quite close to Christopher's hotel, and also not far from the airport. Hmm.

We packed up our cameras. I drove Christopher to the edge of downtown Tucson. I put money in a parking meter. Day 9 was the only day that had taken us to a location with parking meters, and the only day that Christopher had dreamed of a parking meter. We walked about a block, and Christopher noticed a huge tower with numerous satellite dishes. This was the only day where we saw an obvious and unavoidable array of satellite dishes, and the only day that Christopher had dreamed of multiple satellite dishes.

We walked some more, and Christopher began to get frightened. He said he wanted to leave. I asked him why. He reminded me of his dream that later in the day, a woman would be killed not far from where we were walking. I asked, "Shouldn't we check the newspapers to see if the murder has already occurred?" Christopher said, "No. My dreams are quite specific on this point. Let's get out of here."

We returned to Christopher's hotel and, later, drove to the airport, watched some fighter jets land and take off, met Bill, and drove off— three people driving away from an airport, just as predicted.

Next morning, which was Day 10, on the front page of the *Tucson Citizen* was a story of the tragic murder of a seventy-four-year-old woman that occurred less than two miles from where we had been walking. During the ten-day experiment, this was the only day that a murder had been reported on the front page of the Tucson paper— and it was the only day for which Christopher had dreamed of a murder related to the experiment. Hmm.

THE AHA MOMENT—THE CRITICAL TEST OF THE INTELLIGENT DESIGN/HIGHER AUTHORITY HYPOTHESIS

If I recall correctly, it was on the way to the Tucson airport for Bill that I realized the deep significance of the unfolding evidence from the ten-day experiment.

Christopher had insisted from the first that this experiment involved more than just foreseeing the future, but held the potential of revealing the existence of an extraordinary invisible intelligence and power in the universe that played a fundamental role in the conduct of our lives.

If so, then we should be able to discover that the experiment was invisibly and exquisitely designed from Day 1 through Day 10. It should be the case that the shuffling of the envelopes, first by me and then by the secret third person, was actually orchestrated in a way that allowed for exquisite coordination of multiple people's schedules—Christopher's, Bill's, and mine—over the course of the entire experiment.

Then it hit me hard, and I literally could not breathe.

I had not told Christopher that Day 10 turned out to be a special problem for me. In fact, there was no way that I could have known, ahead of time, that there was a potential scheduling problem. It just so happened that Day 10 fell on a Saturday. On this particular Saturday I was scheduled to speak at a conference at the university's medical school. I had to be at the medical school around noon.

I told Christopher this, and said, "If, as you believe, this whole experiment is intelligently designed and orchestrated by a higher authority, then tomorrow's location must be very close to the medical school for me to go with you to the location. However, if you are wrong, and the location turns out to be distant, Bill will have to take you to the site."

(Okay, there are also other alternatives: The hypothesized higher authority might have orchestrated that I would miss the lecture altogether. Or my missing the lecture could have been caused by the fact that I was now worrying about whether God was playing a mediating role in the experiment, and my worrying might, for example, have led me to have a car accident. But an alternative outcome, if it happened, would not serve as convincing evidence that intelligent design was intimately involved in the orchestration of our schedules.)

Christopher's answer shook me to my core—and it still gives me goose bumps today. He said, with his charming British accent, "Gary, no problem. My dream in June says that Day 10 will be over very quickly, and that I will spend most of the day by a swimming pool."

I challenged him because I found it hard to believe. In June, almost two months before I selected the locations or the start date of the experiment, his dreams informed him that Day 10 would be over quickly, and he would spend the day by a pool? Christopher's response was a simple, emphatic "Yes." His dreams at times were fuzzy, symbolic, and suggestive; other times they were clear, literal, and definitive. According to Christopher, this particular June dream was crystal clear, absolutely literal, and completely definitive.

I said, "If your June dream is correct, and the location is truly close to the university, this will provide extraordinary evidence in support of the intelligent design/higher authority hypothesis." Christopher said, "Gary, don't worry. You'll get to your lecture on time. My June dream says so."

I was not worried about getting to the medical school on time. I would be there, one way or another. What I was seriously worried about was the possibility that Christopher might be right, and that the experiment would reveal compelling evidence for the existence of an awesome higher intelligence in the research, and by logical extension, in our personal lives.

I did not sleep well that night. If Christopher's dream, recorded in June, was correct, then scientific integrity would require that I interpret the totality of the data from the experiment, conducted two months later in August, as supporting more than just precognition. The evidence would inexorably point to the active existence of some kind of a higher power orchestration, in a research effort in which I myself was one of the leading actors in an unfolding play.

DAY 10—ARRIVAL OF D DAY

Bill and I arrived at Christopher's hotel at 9 A.M. He reviewed his Day 10 dreams collected from May, June, and July, as well as from the previous night. The information was simple, straightforward, yet surprisingly confusing.

He had seen trees, tall trees, and green grass, as in a London park. He had seen us crossing over a small stream. He had seen an Army ROTC–like building. He repeated what he told me earlier—that this day would end very early, and that he would spend most of the time by the pool.

I heard him say tall trees, green grass like a London Park. I heard him say stepping over a small stream. I heard him say an Army ROTC–like building. I came to the comforting conclusion that only one of the twenty sites fit this description. It was the Summerhaven area at the top of Mount Lemmon. And if that was the location, there was no way that I could take this journey and also be at the medical school on time. The "Guiding-Organizing-Designing" hypothesis would not be supported by the data. I assumed that Christopher must be wrong. I experienced momentary relief.

The Day 10 envelope was finally opened. And I could not believe what I was hearing: the location was not Summer haven but the Arizona State Museum—on the campus of the University of Arizona. The museum was no more than six blocks from Christopher's hotel, and approximately six blocks from the medical school!

I took a slow and deep breath. The proximity of the location fit the intelligent design/higher power hypothesis after all. But I could already tell that for the first time in the ten days, finally, we would find no connection between Christopher's clues and the details of the location.

We packed our cameras. I recalled that the west entrance to the university, the closest entrance to the museum, was under construction. I decided to drive along Sixth Avenue, on the south side of the campus, and pulled into a parking lot a couple of blocks from the museum.

At 9:30 A.M. the three of us were walking behind a row of buildings, including the Social and Behavioral Sciences Building, Centennial Hall, and ultimately the museum. This is the oldest part of the university. As we walked along, I pointed out to the other two something I had never consciously paid attention to before: very tall trees.

Christopher said, "Yes, they are tall, but they are not my trees. These are not the trees I saw in my dreams." We walked some more. I saw more tall trees, and again pointed them out. "Yes," Christopher said. "There are more tall trees here than we have seen at the other nine locations. But these are not my trees, the trees in my dreams."

One of the wonderful things about Christopher is that he is often very conservative about "owning evidence." Having worked with police and intelligence agencies, he knows how important it is to be precise and conservative about this. If the trees don't fit, they don't fit. It's that simple.

Finally, we got close to the back of the museum, and we entered an area that I had forgotten. Here, on a campus in the heart of the Arizona desert, is an area filled with dense trees and green grass. (A conservative estimate is that maybe one one-thousandth of the land area in Tucson has extensive shade trees combined with thick grass, as here.)

Christopher shouted in his exuberant, loose-cannon style, "These are my trees. This looks like a park in London." It turned out this parklike setting was situated next to a Tucson street called Park Avenue; another coincidence?

We continued walking. We reached the construction area, where workmen were hosing down the street. There was a two-foot-wide stream of running water that had to be crossed in order for us to approach the museum. We duly noted the presence of the stream and the need to jump over it. Then we walked up the long set of steps to the museum.

I simply had not thought of the possibility of what happened next: we found that the front door wouldn't open. It was locked. Saturday—because of the construction, the museum was not keeping its usual Saturday hours.

Our destination for the day was the museum; we were here but couldn't go in. It was a little before 10 A.M., and Day 10 was over. Just like that. Finished. The truth is that, scientist or not, I was now shaken.

I said to myself, This can't be true. This can't be happening. Bill was perplexed. Christopher was euphoric. He would spend the rest of Day 10 by the pool, just as his June dream predicted.

But it wasn't over. As we walked back to my car, just before we entered the parking lot, the pièce de résistance occurred. Christopher pointed off to the right and said, "Gary, look at this building." I looked up and saw an Army/Navy/Air Force ROTC building. I shook some more.

We returned to the hotel and reviewed the experimental observations in front of a recording video camera, just as we had for the previous nine days. The recording session ended at 11 A.M. Christopher had not only predicted ten out of ten locations correctly, but his dreams had included detailed information that went beyond the specific locations, addressing the complex coordination of our personal schedules that seemingly defied all conventional reason and logic. I drove to the medical school with plenty of time to ponder what these profound findings meant.

How do we explain these data? Is it possible that an "invisible hand" was playing an incomprehensibly complex role in the orchestration of life—including our personal lives?

In fact, the evidence strongly implied that this superintelligent process was playing a deliberate guiding role in the design and conduct of the ten-day experiment itself.

ARE THESE FINDINGS POSSIBLE?

The truth is, as I review these events today, I still find them hard to believe. However, not only did all these events happen in front of me, they happened *to* me, as well. I was not only an open-minded observer of a complex, carefully designed experiment; I was an unsuspecting participant in a much larger unfolding play—an experimental play involving real science integrated with real life.

It's worth remembering that just because something *seems* impossible it isn't necessarily so. Sometimes science provides compelling evidence that leads us to change our beliefs about what we think is possible.

There was a time, not too long ago, when most of the world's population believed that a round earth was impossible. Scientific evidence now says they were wrong.

There was a time, also not too long ago, when most people believed that a round earth revolving around the sun was impossible. Scientific evidence now says they were wrong.

There was a time in recent history when scientists believed that a massless particle or quantum of organized energy—contemporary physics' concept of the photon—was impossible. Again, scientific evidence now says they were wrong.

Today, we live in a time when many people (particularly in science) believe that an intelligently designed, evolving universe—one reflecting intelligent trial-and-error learning as implied by findings from contemporary evolution—is impossible. However, is new scientific evidence suggesting that we are wrong about this, too?

The late Susy Smith had a phrase for this. The author of thirty books on extraordinary phenomena—including precognitive dreams, survival of consciousness, and the existence of God—she asked me on many occasions whether findings such as these are "too coincidental to be accidental."

Science is a process of discovery, of learning through intelligent trial and error, and of being open for surprises. The fact is, if I hadn't believed in the possibility that Christopher might be right, I would not have been open to seeing the surprising evidence that supported his stunning hypothesis. As Yogi Berra quipped, "If I hadn't believed it, I wouldn't have seen it."

Let's presume, just for the moment, that the findings from this parapsychology experiment, and the subsequent tests I have conducted with Christopher, are valid. Let's presume, just for the moment, that Christopher is a William James "white crow" who disproves the law that "all crows are black." Follow me through some further experiments.

Does the evidence point to the existence of an intelligent, experimenting G.O.D. process and intelligent evolution?

The story continues.

Part Two

SIMPLE G.O.D. EXPERIMENTS

Imagine a set of simple physics and computer modeling experiments you can perform yourself that convincingly demonstrate, time and time again, the inability of coincidence or chance to explain the existence of order and evolution in the universe, including our personal lives.

Is there a plausible, even compelling, scientific reason to believe that the results from these experiments are "too coincidental to be accidental" and that G.O.D. is involved in evolution?

Great discoveries are accidents observed by prepared minds.

LOUIS PASTEUR

3
_____

K.I.S.S.: Keep It Simple Science

SAND PAINTINGS NEVER OCCUR BY CHANCE

We are going to shift gears and move from the parapsychology laboratory to the kitchen and your personal computer. Some of the experiments about which you will read may make you laugh. A few are admittedly far out, but I assure you they all address the existence of a larger spiritual reality—a G.O.D., or Guiding-Organizing-Designing, field—in the universe as well as in our personal lives. Scientists are typically "show me" people who thrive on conducting experiments in order to observe results for themselves. I go beyond that: I am a firm advocate of the K.I.S.S. principle—Keep It Simple Science.

People who have known me over many years will tell you I am a "show me" kind of person who, even as a youngster, had to see it to believe it. Though I was born and raised in the New York area, I somehow identify with the state of Missouri—the "Show Me" state—and I require being shown things in order to believe; my "beliefs" are evidence-based.

There are two "show me" chapters in this book, which include descriptions of a number of very simple yet definitive experiments in connection with the creation and evolution of the universe and of life. These experiments were designed to answer the question of chance versus intelligent design and higher power. You might be motivated to perform these simple experiments yourself, or—even better—with your children. In this complicated world, it's wise to raise children to be skeptical of information tossed at them. We must remember Einstein's wisdom that "the important thing is not to stop questioning."

And so if you are skeptical about any of the information I put forward, I encourage you to perform the experiments yourself. The findings generated in these experiments changed me from the resistant skeptic I once was. In many areas of science, business, law, and everyday life, most evidence allows us to reach conclusions that are "beyond a reasonable doubt." However, the evidence from some of the experiments described in this book, including one seemingly frivolous yet serious experiment I hope you will perform yourself, allows us to draw conclusions that are virtually "beyond *any* doubt."

As you read, I suggest you keep in mind one important question. How can a seemingly frivolous question offer compelling evidence on a subject that has haunted mankind from the earliest days?

THE G.O.D. CONTAINER EXPERIMENT

The story begins almost forty years ago. I was a junior psychology major, chemistry minor, premedical student at Cornell University, on summer vacation in New York City, and I was taken to a small shop in Greenwich Village that specialized in Native American and Eskimo art. It was the first time I saw captivating Native American sand paintings, and I fell in love with them.

I purchased my first exquisite two-foot-by-two-foot sand painting from this shop, and then began to collect others. When I eventually became a professor at Yale, one wall in my study in Guilford, Connecticut, was covered with more than ten sand paintings. My first one still hangs in my Tucson study.

Historically, medicine women and men drew secret sand paintings on the desert floor. The sacred paintings revealed their fundamental spiritual beliefs, and the carefully crafted sacred images were treated with care and awe. Originally sand paintings were made of multicolored grains that were carefully placed on the desert ground. The sands were never glued to the ground—they were intended to disappear with the wind. The images in the paintings available for purchase are intentionally incomplete and inaccurate replicas of the ancient spiritual visions. Obviously, the artworks available for purchase are glued to boards to sustain their fixed patterns.

One day, while I was pondering the origin and evolution of order in the universe, I wondered: if the completely-chance-universe explanation is true, shouldn't it be possible, at least in theory, to create organized, complex sand paintings by simply throwing colored sand in the air?

So I designed and conducted a K.I.S.S.—Keep It Simple Science— sand painting experiment.

I took some white sand, placed it on the bottom of a pot, and then made a simple sand painting with various colored grains of sand. One of the images sometimes painted by Native Americans is frogs. Frogs make me smile. I carefully dripped green and yellow grains of sand and created a mediocre cartoonlike Kermit the Frog sand painting.

I originally used a round metal spaghetti pot with a cover—a square cardboard box with a cover works just as well. Even better is a clear plastic container so you can see the evolving process as it occurs. After my crude sand painting was completed, I covered the pot, and shook it once, opened it, and looked at what had happened to my cartoonlike Kermit. Then I re-covered the pot, shook it again, and took another look. I did this over and over, recording what I saw with each successive shaking. What do you think happened, over and over?

To many people, experiments like my sand painting/shaking experiment might be viewed as a waste since any nincompoop knows in advance what the results will be. So why would a credentialed scientist bother?

The answer is that it is the very nature of science, just as we explain to the children that they are not to accept any conclusion just because it seems obvious. To be accepted, an experiment must be repeatedly performed by reputable scientists who all reach the same conclusion, and that conclusion must become accepted by the community of scientists. If that has not been done, then you perform the experiment yourself. I advise adults of all ages not to accept experimental results on faith despite the frequent temptations to do so. I conduct research, not armchair speculation or wishful thinking. I encourage you to do a few experiments yourself.

What I witnessed that day in the kitchen with my spaghetti pot was truly elementary, and it happened every time. What transpired was that each time I shook the pan, the sand mixed and the frog was no longer.

The more I shook the pan, the more the sand mixed. In the absence of constraints to maintain my original frog design, and in the absence of some sort of a Guiding, Organizing Designer taking a hand, the sand participated in a most remarkable and beautiful process. The process is simple and completely replicable. The sand mixed. Blacks and whites, yellows and greens, whatever colors of sand were present, they were all brought together. They became a blended mixture, a family of colors, so to speak. They became a complex yet fairly uniform (more on this later) mixture of different colors.

Being a "show me" scientist, I insisted on replicating this experiment many times. Sometimes I began by creating an image of a frog, sometimes I began by writing the word "frog." Sometimes I drew a picture of a heart, sometimes I began by writing the word "heart"—it didn't matter. When I went through the successive shaking of the pot, the result was always a blended mixture. Always. The sand always mixed, period. The conclusion is inexorable and unstoppable.

In the absence of some sort of Guiding-Organizing-Designing process, sand mixes. It mixes every single time. Chance by itself does not create sand paintings. When given a shake, sand mixes. When mixed by wind, the sand images disappear and blow away.

Though this outcome may seem obvious to you, I want to illustrate the significant take-home message of this experiment through my direct personal experience.

In my childhood and adolescence, over a time period of more than ten years, I walked upon thousands of miles of beaches. Growing up on the south shore of Long Island, I loved the ocean. In all that time I never came upon a frog-shaped sand painting, or any other Native American painting made of sand. I have looked carefully at thousands of pictures of beaches from around the world; I never discovered a single sand painting. I have walked upon thousands of miles of desert, and never observed a single sand painting.

In my university lectures, I have asked my undergraduate students at Harvard, Yale, and the University of Arizona if they have ever chanced upon a sand painting, either on beaches or in the desert. Once again, not one ever reported having seen spontaneous sand paintings.

If "randomness" could indeed create sand paintings on beaches or on deserts—as the completely-chance-universe hypothesis predicts—how come no one has ever reported seeing a spontaneously created sand painting? Where are the sand paintings?

If chance cannot create simple sand paintings, then how can chance create an entire universe that is organized and evolving? It isn't logical to attribute profound organizing power to chance in the absence of any positive evidence. The conclusion of the sand painting experiment for me is quite forthright and even simple—and it is definitely one worth teaching to our children. It is no more probable that sand paintings will spontaneously, by chance, appear on a beach or in a desert than that they will spontaneously, by chance, appear in my spaghetti pot—or in yours.

This is a fact. Sand paintings never spontaneously occur, either in pots and pans or on beaches and in deserts. Again, if you are of a highly skeptical nature and unwilling to accept the logic of this, do the pots and pans experiment yourself.

If you happen to come upon a detailed, multicolored sand painting similar to the intricate two-foot-by-two-foot sand painting I purchased so many years ago in that little Village shop, you will know for certain that some highly intelligent Guiding, Organizing Designer—or intelligence of another sort, such as a Native American artist—carefully put it there. When you fully understand the sand painting experiment and its conclusions, you can know without any doubt that it was not created by chance.

FROM SAND PAINTINGS TO COMPLEX COMPUTER SOFTWARE AND DNA

If sand cannot arrange itself by chance, what do you think the probability is of getting the current complex version of the Windows operating system to spontaneously appear, by chance, in a PC, or a hugely more complex DNA molecule to spontaneously form, by chance, in a cell?

If, in the absence of a highly intelligent Guiding-Organizing-Designing process, what nature does is mix things, then what do you think will happen if we use the numbers generated by carefully measuring the mixing of black and white grains of sand, and use these resulting patterns of numbers to organize bits of 0's and 1's in your PC? Or if we use these patterns of numbers to organize the four bases that make up DNA molecules in your cells? Will we ever get a working computer operating system, or a functioning human being?

If the completely-chance-created-universe hypothesis predicts that working operating systems or organisms will appear by accident, yet simple experiments designed to test the hypothesis reveal that, like sand paintings, real-life working operating systems or organisms never spontaneously appear by chance, then something is seriously wrong with any current explanation of how chance works.

Chance is not a highly efficient way to create useful computer programs or for nature to create living organisms. In truth, it's an impossible way.

SEARCHING FOR THE CORRECT EXPLANATION

So is there an explanation for the fact that millions or billions of monkeys, given millions or billions of years, could not create your computer's operating systems or your personal DNA, let alone create a sand painting by throwing sand in the air?

When chance alone cannot even write the word "chance" in the sand, how can a small group of vocal scientists and skeptics, like Richard Dawkins, Ph.D., and Michael Shermer, Ph.D., claim that the profound organization and complexity observed in the universe can occur by chance? These people claim that chance played a critical role in creating everything from subatomic microstructures to superclusters of galaxies, and all things in between, including humans. Why? Simply assuming that after trillions and trillions of random trials, order in the universe can occur by chance does not fit with the obvious experimental evidence indicating that carefully created sand paintings, when allowed to do so, mix 100 percent of the time and are never replaced by new sand paintings.

In fairness to mainstream scientists, the generally accepted conceptual model they employ to explain the origin and evolution of order in the universe is a bit more complex: the model assumes that chance is combined with physical laws and chemical laws that limit and guide the ordering process. For example, Dawkins writes in his book *The Blind Watchmaker* that evolution is a gradual, step-by-step process of transformation "from simple beginnings, from primordial entities sufficiently simple to have come into existence by chance." He further concludes that the subsequent evolution of biological organisms is presumed to be guided by "natural selection" and "self-organization," combined with "random mutations and variations" of DNA. However, their bottom line is still that the foundation of the evolution of the universe implies the operation of chance, since it presumes intelligence was *not* involved in the creation and organization of the fundamental physical and chemical laws that supposedly limit and guide the creation and organization of "primordial entities" as well as complex material systems—both nonliving and living.

I do not wish to criticize either my former teachers or my contemporary colleagues. And I certainly do not intend to be mean-spirited. Until I saw the lesson of the sand painting experiments, I perceived things just as Dawkins and Shermer did. I was one of them.

I was once a young scientist who learned the well-accepted explanations and predictions regarding the primacy of chance and randomness from my distinguished Ivy League professors, and then passed along their seemingly logical conclusions to my students. Honestly and with integrity, I gave my students the explanations my professors gave to me. When I began teaching at Harvard, I did not know that some of these fundamental explanations were seriously in error, and that I was innocently and inadvertently promoting mistakenly accepted knowledge.

However, after my sand painting experiments at Yale, I realized that there was something seriously wrong with the foundation of the underlying logic of the chance-universe explanation—because the logic I was taught simply didn't match the experimental evidence. The theory didn't fit the facts. The reasoning did not fit the observations. I came to realize that something somewhere was clearly in error. And I again began to question the conclusions of the scientists who had been my wonderful mentors.

IS THE WATCHMAKER BLIND, OR IS IT WE WHO ARE BLIND TO THE WATCHMAKER?

Does unbridled chance really occur in the universe? Can the evolving universe as we experience it at this moment in history, including the laws of physics and chemistry, really have occurred by chance?

Or is it necessary for us to logically deduce the existence of some sort of superintelligent Guiding-Organizing-Designing process at every level of nature, from the micro to the macro? Do separate Guiding-Organizing-Designing processes exist in nature, or are they all an expression of one universal G.O.D. process? Is this what is implicitly meant by the Great Spirit, the Living Source, or simply, "God"? Is this what we mean by higher purpose, higher intelligence, and higher power? Is this what a larger spiritual reality means?

When I met Christopher Robinson and saw how far the "too coincidental to be accidental" process could be taken—literally, all the way to the orchestrated coordination of our personal lives, and beyond—I was shaken to my core.

As you will read in Part Four of this book, the explanation for the existence of some sort of an intelligent G.O.D. process is actually easier to grasp than it is to learn how to make a complex Native American sand painting, or for that matter to even learn how to fold your shirts.

As everyone knows, throwing ten shirts in the air *never* results in the shirts spontaneously coming down neatly folded, one shirt on top of the other. I underscore "never." If you doubt this, do the experiment. Being a "show me" scientist, I tested the hypothesis and actually threw my shirts in the air. What came down, every time, was more or less a mess. Actually it was always a mess, a big one.

If you still doubt the generality of this conclusion, take Lincoln Logs and throw them in the air. Do they ever spontaneously make a log cabin? The answer is: never. Try throwing the pieces of a Lego set in the air. Do they ever come down and spontaneously make a dump truck, or a robot? The answer is: never. Take the pieces of a mechanical watch—the gears and springs and screws and hands—and throw the pieces in the air. Do they ever come down assembled as a working timepiece? The answer is: never.

If the intelligent G.O.D. process explanation is correct, then Dawkins's universal watchmaker—from his book *The Blind Watchmaker*—is not blind at all. In his book Dawkins, who is Oxford-based, describes evolution as a blind, nonintelligent, yet creative (as in *unpredictable*) process where predefined designs are found. The subtitle of his book sets forth his "nonintelligent" theory: *Why the Evidence of Evolution Reveals a Universe Without Design.*

But despite being winner of a Los Angeles Times Book Prize and receiving the Royal Society of Literature's Heinemann Prize, Dawkins got it wrong; his conclusions, though beautifully written, are backwards. It is we who are blind to the watchmaker.

It's time for those who adhere to the chance theory to remove their blinders, open their eyes, and celebrate the exquisite ability of humans to observe purpose, intelligence, power, order, and evolution in the universe.

And if you are wondering about so-called natural selection and self-organization in nature, I encourage you to keep reading—and be prepared for a paradigm-changing surprise.

The final story of randomness—utter chaos—has not yet been told to us by the mathematicians. It seems remarkable that something so fundamental for probability theory has not been defined and even more remarkable that we can go so far in mathematics lacking a definition. By simply assuming randomness exists, mathematicians assign elementary probabilities to events, and that is their starting point. But they have not captured chaos and looked it in the eye.

HEINZ PAGELS

4

G.O.D. in the Computer

HOW YOUR PERSONAL COMPUTER CAN
TEACH YOU ABOUT GOD

Home computers operating at gigahertz processing speeds are performing billions of calculations per second. It is possible to conduct complex mathematical experiments on personal computers and process the data more rapidly than ever before. Sometimes the evidence from these modeling experiments is so replicable that a researcher can draw conclusions that are virtually beyond any doubt.

And now I will share with you how your personal computer can be used to support the discovery that chance is not a plausible explanation for the origin and evolution of complex orders in the universe. By complex orders I mean, for example, the patterns of billions of chemicals called bases organized in DNA, or the patterns of billions of stars organized in spiral galaxies.

As you will soon see, you can use a home computer to conclude definitively that even randomness itself does not occur by chance. And if randomness does not occur by chance, then it's time for us to understand these three things: (1) why randomness is incorrectly described as "disorder" or "complete unpredictability," (2) how randomness really functions, and (3) what role it plays in the intelligent G.O.D. process universe.

In the following pages, we will be looking randomness "in the eye," as the distinguished physicist Heinz Pagels put it. Perhaps, if I am convincing enough, you will have the courage to see randomness and chance from an entirely new vantage point. The computer experiments I will describe use simple arithmetic—nothing more than addition and division—and are very easy to understand. I promise you will be able to follow the reasoning even if you are shy of math or science.

People turn to the computer to take the drudgery out of calculating the millions of simple additions and divisions that until the middle of the twentieth century had to be done slowly and painstakingly with pencil and paper. Contemporary digital calculators and computers perform arithmetic operations with absolute accuracy and in less time than it takes for a heart to beat.

I am about to show you that when numbers are arranged with the conditions necessary for "randomness" to occur, we see a predictable *order* every single time. The order—actually termed a *random* order—is absolutely normal and is not random at all.

In other words, when the precise conditions for producing randomness are created in a computer, a completely normal and 100 percent replicable order emerges that can be seen with the naked eye.

Now is the time to see randomness "with new eyes," as Marcel Proust would say, and have fun in the process. (But if you are still, despite my reassurances, put off by computers and arithmetic, you can decide to accept this conclusion as true, skip this section, and jump ahead to Part Three.)

HOW RANDOMNESS CREATES NORMAL DISTRIBUTIONS

There are many computer operations in which we need to have a number—or many numbers—selected at random. One leading example is encryption programs, including the kind that protect your credit card number as it speeds along the Web to the online merchant you're buying from. Computers can select numbers randomly using so-called random number generators. The computer can select numbers in a seemingly unpredictable manner.

For example, if we instruct the computer to pick a number between 1 and 100, there is no way to predict ahead of time the precise number it will pick. The computer is just as likely to pick 1, 2, 3, or 4 as it is to pick 97, 98, 99, or 100.

If the computer happens to pick the number 1, for example, and then we ask the computer to pick a number from 1 to 100 again, it is just as likely to pick the number 1 again as it is to pick any of the other ninety-nine possible numbers. This is because each calculation is set up to be independent of the previous calculation. The process of selecting any number from 1 to 100 can be performed hundreds, thousands, or millions of times—almost completely independently and unpredictably.

I say "almost" because digital computers only approximate true randomness. Theoretically, true randomness—more precisely, what is termed random sampling—requires complete independence from calculation to calculation. If each and every calculation is not performed completely independently, true randomness cannot occur.

Suspend belief for the moment to consider that the computer approximates complete independence and say that you can program the computer to select a range of possible numbers completely randomly. Let's say you ask the computer to select a number from 1 to 100 randomly, and to do that a hundred times in a row.

The computer can be programmed to calculate the average of these randomly selected numbers (add the hundred selections together and divide the sum by 100), and to save the resulting average. And you can program the computer to repeat this process over and over.

It's important to understand why we are calculating the average of the randomly selected numbers in order to discover the true order in randomness. Since the computer in this experiment is programmed to select the numbers from 1 to 100 randomly, it is extremely improbable that it will select, by chance, any single number 100 percent of the time. Instead, out of a hundred numbers selected, approximately half the time it will select numbers that are less than 50, and the other half of the time it will select numbers that are greater than 50. This means that the average of the hundred selections should tend to be around 50. If you have any background in math or the sciences, you're still with me. And even if you don't, I hope you'll make the effort to follow the reasoning.

Now, since the selection process is random, sometimes there will be more numbers selected that are above 50. This will tend to push the average of the hundred selections above 50. Sometimes there will be more numbers selected that are below 50. This will tend to push the average of the hundred selections below 50. However, we can expect, by chance, that the average of a hundred random selections will most frequently be around 50.

An average in any one run of, say, 45 or 49, or 51 or 55, isn't particularly unusual. But as you move further away from 50 in either direction, there is less and less chance of that average number occurring. In the two experiments performed and graphed below, none of the averages reached below 40 or above 60.

Notice that the average of a hundred randomly selected numbers will almost never be 1, because this would mean that the computer had selected the number 1 every single time. Conversely, the average of a hundred randomly selected numbers will almost never be 100, because this would mean that the computer had selected the number 100 every single time.

In the two experiments I present below, I asked the computer to (1) select randomly any number from 1 to 100, (2) repeat the selection a hundred times, (3) calculate the average of these hundred selections, (4) save the average, (5) repeat these four steps a hundred times, and (6) plot the results.

Distribution of 100 averages of random numbers from 0 to 100

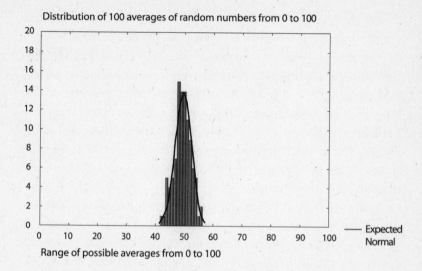

Chart 1. Bell-shaped, "normal" distribution of a hundred averages of sets of one hundred numbers randomly selected from 1 to 100.

Chart 1 plots the distribution of the one hundred averages that were calculated and saved by the computer. These are real averages. The possible averages, ranging from 1 to 100 in steps of 2, are indicated on the horizontal axis. The number of averages that were actually calculated between 48 and 50, 50 and 52, 52 and 54, and so forth (termed "number of observations"), is plotted as bars.

When the computer generated the one hundred separate averages, and then calculated the grand average of the averages—in other words, the average of all of them—the grand average turned out to be 49.9898, which is virtually 50.

The lowest single average was 44.5512, and the highest single average was 57.0632. The distribution of averages approximates what is called a bell curve—with a bell shape that mathematicians and scientists refer to as a "normal" curve.

Notice that the actual distribution is not a perfect bell-shaped curve, but *approximates* a bell-shaped curve. If instead of performing only a hundred averages, we had performed five hundred averages, the resulting curve would have looked more perfect.

For comparison, I had the computer repeat the experiment and plot the findings from a replication experiment; see Chart 2.

The new graph is virtually the same as the previous one. The precise distribution of the bars has changed somewhat (the tallest peak is at 54 for Chart 1, and 46 for Chart 2), but the pattern is very similar. When the computer generated the average of the averages for this experiment, the grand average of all of them was 49.8558—again virtually 50. The lowest single average this time was 42.9938, and the highest single average was 56.4059. The distribution again approximates the bell-shaped curve.

Consider: If we have the computer repeat this process, again and again, what do you think we will observe? Will we ever see a pattern of bars that averages 30, with a range from 20 to 60? Or will we ever see a pattern of bars that averages 70, with a range from 50 to 90? As you probably guess or assume, the answer to that question turns out to be no. What we will see, more or less precisely, is what I have twice graphed here. The average turns out to be virtually 50, with a range from 42 to

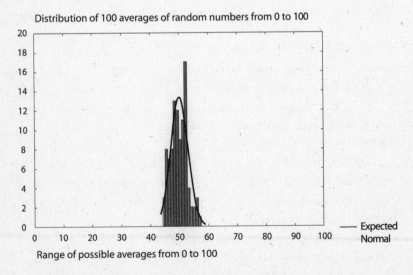

Chart 2. Results of a second test of a hundred averages of sets of one hundred numbers randomly selected from 1 to 100.

58. Numbers selected entirely at random, if you select enough of them, always have an average that is exactly in the middle of the range.

Here again, you need not take what I'm saying on faith. I certainly did not come to this conclusion on faith. I insisted upon seeing experimental evidence with my own eyes. You're welcome to do the same. Only after repeating this experiment hundreds of times did I come to evidence-based faith in the complete replicability of randomness.

I originally did this experiment as a professor at Yale University over twenty years ago, using an Apple II computer with a miniscule (by today's standards) 64,000 bytes of memory. I programmed the arithmetic using an elementary computer language called BASIC. I had the computer automatically plot the resulting distribution of bars all day long, and print the results to my Epson dot matrix computer. Graphs 1 and 2 were generated at the University of Arizona a few years ago on a Hewlett-Packard Pentium III PC with 128 million bytes of memory. I used Statistica for Windows software to perform the arithmetic. The computer took but a few seconds to perform all the calculations for a given experiment and plot the graph on a high-speed laser printer. Were I to repeat the calculations now, I would be using my current office machine, a Dell Pentium IV PC with 512 million bytes of memory. The calculations for a given experiment would take but a fraction of a second.

However, whether I performed the calculations for a given experiment by hand (which would take many hundreds of hours), or with my old Apple II computer (which took approximately ten minutes), or using my blazingly fast new desktop computer (which would calculate almost as fast as I can hit the Enter key), the pattern of results wouldn't change—because the findings are 100 percent replicable: The average of the bars per experiment will be approximately 50. *Every single time.* In other words, since randomness was generated mathematically, and the experimental conditions allow randomness to occur every time, an overall normal distribution was obtained *every time.* Moreover, the pattern of the bars was always bell-shaped. The pattern was never square, or oval, or inverted, or a straight line. The pattern was a bell-shaped order, period.

So what does this mean?

When one purposely sets up the precise conditions that allow randomness to occur, in the absence of a specific Guiding-Organizing-Designing process—for example, if there is no intervention by the invisible hand of a G.O.D. process—what we observe is a completely replicated random order which is expressed as a bell-shaped curve. The appearance of a bell-shaped curve by a random-number-generating computer is as replicable as throwing your laundry up in the air and having it come down in a messy, unfolded pile.

Notice that this computer program never generates an average of the averages that would indicate that the same few numbers were picked again and again, or that the same number was picked every single time. An average of 5 would require that most of the numbers selected were below 10; an average of 95 would require that most of the numbers selected were above 90. An average of 1, for example, or an average of 100 would require every single number selected to be a 1 in the former case or 100 in the latter.

So if the circumstances are set up so that numbers are selected completely randomly, the resulting pattern always averages out to the bell-shaped curve. And the more observations that we make, the more perfect the bell-shaped curve becomes.

If we accept the evidence, the conclusion is obvious. In the absence of a Guiding-Organizing-Designing process, the material world mixes (Chapter 3) and the mathematical world creates a normal distribution, as we have just seen.

In a word, there is something profoundly orderly about randomness. When we design the computer modeling experiments properly, we can even see this order with the naked eye.

DEEP QUESTION: HOW DO WE KNOW IF A SEQUENCE IS RANDOM?

In Heinz Pagels's brilliant book *The Cosmic Code: Quantum Physics as the Language of Nature,* the chapter titled "Randomness" led me to realize something profoundly disturbing about our arrogant tendency to label sequences of numbers we cannot understand as random or disordered.

Pagels, a gifted physicist, was trained at Princeton University and served as a professor at Rockefeller University. In his chapter on randomness, he explains how randomness is an interpretation that seems to be true but may in fact not be. In one place, he writes:

> *This example illustrates that what we may think is a random number really isn't—it is related to other numbers which are specified by a simple rule. How can you be sure a number is truly random? You can't— the most you can do is establish the number is not random if it fails one test for randomness.*

And elsewhere, equally wise and empirically correct:

> *A remarkable feature of "random" numbers—numbers that pass all the tests—is that two such numbers may be related to one another in a nonrandom way.*

What this means is that even though the complete series of numbers may appear to pass all known mathematical tests for randomness, subsets of the numbers in the complete series can be easily discovered to be nonrandom.

How do we know for sure, then, if the series of numbers is random?

Let's consider something we are all familiar with—music—and see what we can learn about whether music is random.

SING A SONG OF ORDER AND RANDOMNESS

While still in high school and college, I worked as a professional musician and considered a career in music. So it may be natural for me to identify a connection between music and the question of randomness.

We begin with a question: how is it determined if a given sequence of numbers is random?

Chart 3 shows a combination of numbers from 1 to 6 that are

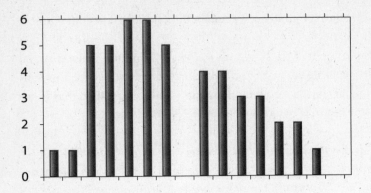

Chart 3. Numbers from 1 to 6 arranged in a particular sequence.

plotted as a sequence of bars over time. I remember feeling delighted when the insight underlying this curve first came to me.

When I teach this principle to students, I usually plot this sequence on the blackboard and ask the students what it looks like. Some say it looks like the New York City skyline. Some say it's shaped like a mountain outside Tucson. My physiology colleagues say it's shaped like a sweat-gland response. People sometimes say that, tilted on its side, it looks like a nose or a female breast.

It is certainly a familiar shape. But where precisely did the sequence of numbers 1155665044332210 come from? Is it random?

The answer is, this is the actual sequence of numbers that specifies the tune of the children's song "Twinkle, twinkle, little star, how I wonder what you are."

Everyone knows that the "Twinkle, twinkle" melody is a simple, ordered set of notes. Its temporal sequence—that is, the order looked at on a time basis, one note following another—is obviously not disordered. It would fail very simple tests of randomness.

Now, imagine what would happen if we changed the order of the notes? Would the "Twinkle, twinkle" melody disappear? Of course. However, would it become another melody? Depends upon what you consider to be melody.

Let's take the "Twinkle, twinkle" numbers and analyze them as we did the random selection of numbers at the beginning of this chapter. When I place the 1155665044332210 sequence of numbers into my statistics package, and display the numbers in terms of their average and histogram (distribution of numbers), Chart 4 shows what I get.

First, when we mix the sequence of numbers and calculate the average, we completely lose the melody. When we calculate an average and plot the distribution, we lose the temporal pattern—the melody, the order over time—that defines "Twinkle, twinkle."

What we learn is that "Twinkle, twinkle" has a mean of 3.0000 whose lowest value is 0 and whose highest value is 6. There are two 0's, three 1's, two 3's, two 4's, three 5's, and two 6's. If the distribution were normal, the curve would look bell-shaped, as the expected normal shows. Obviously this distribution of numbers does not look normal.

Now, what happens if we change the order of the "Twinkle, twinkle" numbers? Instead of allowing it to be 1155665044332210, we allow it to be 5136040521623541. This sequence looks random. And it

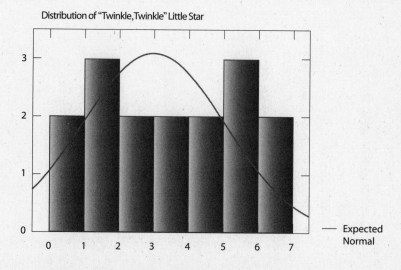

Distribution of "Twinkle, Twinkle" Little Star

— Expected Normal

Chart 4. Distribution of "Twinkle, twinkle" numbers.

certainly does not sound like music to our ears—at least as we define music. The plot is shown in Chart 5. When you look at the pattern, it's obviously very different from "Twinkle, twinkle." Remember, this graph contains the identical number of integers; only the arrangement of the sequence of the integers over time is different.

Now, what happens when we obtain the average and distribution of these numbers? What do we see?

What we see, obviously, is the identical average and distribution as before—as in Chart 4. This is because when we calculate the average and the distribution of the integers, the result is not influenced by the temporal order that may be in the data.

In other words, just because the mean of two distributions may be precisely 3.0000 and have the same shape, this does not mean that the temporal order of the actual numbers comprising the sequence is identical. In fact, temporal order gets lost when we analyze the data this way.

For the record, I do not want to lose the music—I want to discover the melodies and harmonies and celebrate them. I still think of myself as a musician, and therefore I am very sensitive to sequences of notes/numbers over time. If the music is there, I want to hear it. I want to hum the melody. And even dance to it if possible.

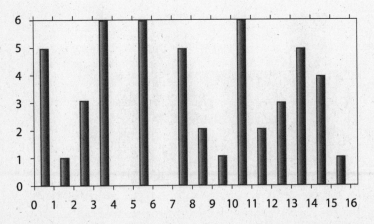

Chart 5. Numbers 1 to 6 from Chart 3, arranged as a different sequence over time.

TWINKLE, TWINKLE, LITTLE PI?—IS PI TRULY RANDOM?

One of the great unsolved mysteries in contemporary mathematics is the sequence of digits of the famous (and some say sacred) number pi.

Pi is a remarkable number. It is, on the one hand, so simple—it is the circumference of a circle divided by its radius. One might think that it should be easy to calculate the number pi, but in reality this is not true. Supercomputers have been used to calculate the division of the circumference of a circle by its radius to *hundreds of thousands* of digits, and no one has been able to discover what the precise order is. No one has discovered a replicable pattern to the sequence. The pattern of integers appears random to the most stringent mathematical tests.

And yet, the precise sequence of pi's numbers is 100 percent replicable. Each and every time a mathematician computes the division comprising pi (using a given calculation procedure), the sequence of numbers comes out to be identical. The sequence is replicated, over and over.

Of course, the sequence of numbers when 7 is divided by 4, or 11 is divided by 3, is also 100 percent replicable. However, with these ratios, it is possible for mathematicians to discover that the pattern or sequence of the resulting integers is not random. What is special about ratios such as pi is that its pattern or sequence of resulting integers has defied the brightest mathematicians and supercomputers from discovering any apparent order to the sequence.

When scientists observe a pattern of data that replicates over and over, they come to the conclusion that the probability of obtaining such data by chance is virtually impossible. How can such a random order occur by chance if the order emerges the same way every time and therefore defies chance?

There are two choices facing us—we can conclude that patterns can be random even though they are replicable, or we can conclude that what we typically label as randomness can sometimes reflect a level of organized complexity that is beyond our current ability to describe mathematically.

When patterns in nature are discovered that replicate over and over, scientists slowly but surely persist until they come to understand the rules that govern the order. For example, when Newton observed that little apples and big apples fall to the ground at the same rate, he persisted until he could arrive at a conclusion about the nature of gravity.

SUMMING UP

What has been revealed about randomness in this chapter?

That randomness does not occur by chance. And that randomness produced by random sampling requires events to be completely independent of one another. We have learned that when events are independent—as modeled by random number generators in modern computers—the resulting distributions of averaged numbers are anything but random. We also learned that the selection process creates a precisely organized distribution of averages that is considered to be the normal, bell-shaped curve.

The evidence of a replicated bell-shaped curve demonstrates that all potential orders do not spontaneously occur when randomness is allowed to operate blindly. Simply stated, the empirical evidence from experimental computer modeling is inconsistent with the common-sense prediction.

By inferring randomness from the temporal (time-based) sequence—as illustrated by "Twinkle, twinkle, little star"—it became clear that sequences of numbers can reflect intelligently created patterns. I believe it's vital that scientists learn humility when patterns and sequences appear too complex for them to be specified mathematically.

We also learned in this chapter that statistical methods used to calculate averages and distributions, though helpful, pay a severe price when it comes to revealing ordered patterns of numbers over time. In particular, the mysterious number pi is especially telling because its complex sequence of numbers defies description (and is therefore labeled as random or disordered) even though the pattern of digits (the *order*) is completely replicated each time it is calculated.

So we've seen in the previous chapter that sand does not spontaneously make sand paintings—it mixes—and in this chapter that randomness does not make multiple melodies: it makes bell-shaped curves.

WHAT THIS MEANS

Order does not occur by chance—but neither does randomness. The logic becomes inexorable. The conclusion becomes inescapable. If complex orders do not occur by chance—the clear take-home message from this chapter—and we discover replicable evidence of complex orders (be they in sand paintings or in sequences of numbers we experience as melodies and harmonies), then we can't logically draw the conclusion that the replicated orders could have occurred by chance alone. *Chance per se is no longer a plausible explanation for the existence of order.* It's that simple.

Since we have ruled out the "chance" explanation, we therefore must consider an alternative. When I learned what you now know, I proposed that the most plausible alternative explanation for the incontrovertible evidence of the existence of complex and evolving orders in nature and the universe—whether the process is small or large, visible or invisible—is the presence of what I have come to call the Guiding-Organizing-Designing process. *In the language of physics, this implies the existence of a universal "field"—including gravitational and magnetic fields—that functions as a formative process and provides intelligent guidance, organization, and creative design.*

Given this logic, I wondered what would happen if I allowed myself to accept the fact that some sort of universal G.O.D. process exists in the universe?

What would happen if I suspended my doubt and requested information from the G.O.D. process? What if I asked the G.O.D. process from the bottom of my heart as I ask for help from my closest friends?

Could I contact G.O.D.?

Could we communicate?

Could I understand the communication?

And could I ultimately build a relationship? The experiments now become highly personal.

Part Three

CAN YOU COMMUNICATE WITH G.O.D.?

A reluctant scientist asks the universe a seemingly trivial question and receives a superficially silly but ultimately profound answer that shakes him to his core. The scientist is me, and I repeat this exploratory experiment many times, with equally startling results.

The observations suggest that suspending our doubt with open hearts can allow us to receive G.O.D. information that is evidential and at the same time amusing.

To see or not to see, that is the question . . .

Ask, and ye shall receive.

JOHN 16:24

5

Talking to an Intelligent Black Box

HOW SCIENCE FICTION SOMETIMES PROVIDES A KEY TO REALITY

Is it possible that you and I can learn to communicate consciously with the G.O.D. process? And may we be able to teach others to do the same?

Since each of us is an intrinsic part of the universe, aren't we all part of the G.O.D. process? Since I am personally part of the process— as are you—is it possible that I can personally discover what this intelligent process is, at least to some degree?

I can't know unless I ask. And you can't know until you ask, either. We must each be willing to ask if we are to receive.

A WORD OF CAUTION

The problem is, it's easy to fool ourselves. In fact, as a species we are masters at this. Psychologists are trained to understand the many tricks the human mind can play on itself, including the creation of:

Illusory correlates—when we imagine relationships that do not exist

Self-deceptions—when we deceive ourselves and ignore evidence that contradicts our beliefs

Confirmation biases—when we mistakenly find support for the things we believe must be true

False memories—when we inaccurately remember details of something that happened, or think we remember something that did not actually happen

The list of what psychologists call "cognitive distortions" is extensive and the research literature keeps expanding within the areas of cognitive and social psychology.

Not surprisingly, psychologists are among the most conservative of scientists when it comes to openness about what might be termed "anomalous" communication—meaning communication that deviates from what is believed to be possible or what is expected. This is especially true of purported communication with spirits—precisely because scientists know how easy it is to be fooled by one's own mind.

And the same holds true for the highly skeptical attitude of many psychologists toward the G.O.D. process.

I was fortunate to have received excellent training in cognitive biases and distortions in my graduate education as a research psychologist, including an especially valuable graduate course in psychopathology with Professor Loren Chapman. It was Chapman who, while at the University of Wisconsin, discovered and coined the term "illusory correlate phenomenon." And I was blessed to have Professor Robert Rosenthal, the father of research on "experimenter effects" (inadvertent experimenter errors), on my dissertation committee at Harvard.

For these reasons, I'm vastly more mindful than the average person, and even considerably more aware than the average scientist, when it comes to understanding the need to be extremely cautious about collecting and interpreting psychological data, especially data provided from one's own mind and from one's own subjective experience.

I was also proficiently trained as a clinical psychologist at Harvard. I learned about schizophrenia and hallucinations, mania and delusions, multiple-personality disorder, and dissociation. I deeply respect the pressing need to be vigilant for the possibility of psychopathology, especially when it comes to the claim that someone, including myself, is having "conversations with the universe."

So why am I about to commit to print some of my initial personal experiments in attempting to communicate with the hypothesized G.O.D. process?

There are three reasons:

Reason 1. As indicated in places earlier in this book, there is good scientific reason for predicting that communicating with a G.O.D. process is plausible. If it is theoretically plausible, *and* if it is supported by experimental evidence, it is important to acknowledge these facts.

Reason 2. As long as the information obtained is received with caution and humility—with full awareness of the limitations of my mind, including my inherent capacity to fool myself—it is possible, in principle, to obtain meaningful and even accurate information.

Reason 3. Evidence for a genuine spiritual reality is currently being observed and documented by contemporary research on survival of consciousness. Some of these findings are covered in my previous books, *The Afterlife Experiments* and *The Truth about "Medium."* So, in principle it seems, all people have the potential to explore a larger spiritual reality. Therefore as a professional scientist I am obliged to be among the pioneers and the explorers. It is always valuable for a scientist to be a subject in his own research. Ultimately there is no substitute for direct experience.

Before describing my efforts, I first want to review the plausibility of communicating with "the Source," and to show the way systems science predicts not only that this is plausible but that it is probable.

IS THE "ALL" WITHIN THE "SMALL"?—AN INTRODUCTION
TO SYSTEMS SCIENCE

The branch of science known as systems science teaches that there is a universal principle of the micro (the little) being contained within the macro (the big). This is because systems are composed of smaller parts (termed subsystems) and are simultaneously components of larger units (suprasystems).

For example, in multicelled organisms (which are one type of a system), individual cells (tiny subsystems) are contained within the body (which compared in size to individual cells can be huge, especially when the body contains trillions of cells). And individual organisms are part of larger suprasystems, from families to ecosystems and beyond.

The subsystem/system/suprasystem organizational structure extends from subatomic particles to the universe as a whole. Stated simply, the Small is in the All. However, systems science also teaches us the less obvious fact that the reverse can also be true—the All is in the Small. This is not only semantically true—the word "all" is literally in "small"—but it is organizationally and empirically true, too.

In the case of the body, for example, every cell is nourished by the blood; the energy, and all the chemicals and other materials within the body, are mixed within this living liquid to various degrees. And this goes further; it's not just materials and energy that are carried by the blood. Bioelectromagnetic information and energy signals—generated by every cell and carried by bioelectromagnetic fields—also circulate in the bloodstream, by means of what is termed "volume conduction." The resulting info-energy-material mixture, the "whole" of the blood—which contains the All of the body—is continuously circulated back to every cell within the body. This realization is similar to the concept of the hologram, where the All (the holo) is optically embedded within the Small.

The word "mixture" here may remind you of our earlier discussion in Part II. Recall what happens when a selective organizing process is *not* operative, and things simply mix. In physics this is called the Second Law of Thermodynamics. Systems science suggests the possibility that a higher purpose exists for the second law in an intelligently designed, evolving universe: that it has the purpose of fostering a universal mixing process which operates at every level. And the reasoning is straightforward. Universal mixing insures that sharing, interconnecting, and creating of wholes occurs at all levels in the universe. Simply stated, universal mixing insures not only that the All will continue to exist (in the Small) at all levels, but that it will continue to evolve at all levels as well.

"AS BELOW, SO ABOVE"

Another well-known phrase, "As above, so below," proposes that patterns existing at macro levels can also be observed at more micro levels. In religious terms this is usually interpreted to mean that what happens in "heaven" can happen on earth, a supposed phenomenon sometimes called "heaven on earth."

What systems science teaches is that the reverse can also be true: "As below, so above." When I first read James G. Miller's book *Living Systems,* I was led to ponder the intimate relationship between life on earth (the Small and below) and the universe as a whole (the All and above).

However, even after my first attempt to conduct a personal exploratory experiment testing the All-in-the-Small hypothesis by attempting to communicate with the universe (confessed in the next chapter), I didn't allow my mind to explore this possibility seriously until I moved to the University of Arizona. There I would soon meet a courageous undergraduate student, Sabrina Geoffrion, who saw a significant connection between my explanation of systems science and her personal vision of an evolving Source in an evolving spiritual universe.

Sabrina had been encouraged to think big. At that time both her parents worked at the University of Arizona—her mother as professor and head of the art department, her father as associate vice president for research. Though her mother and father were academically conservative, they were intellectually open and loving people. They encouraged Sabrina to fearlessly and openly explore her own mind and they inspired her to dream.

As a result, Sabrina put forth the possibilities based on this starting point: Assume systems science predicts that what we are calling the G.O.D. process is the macro system. Then according to contemporary physics, all people would be intimately connected with the macro system (that is, "God") to varying degrees. The logic comes from basic physics. All physical objects—from atoms and chemicals, through organisms and planets, including solar systems and galaxies—are interconnected in various degrees by electromagnetic fields (as well as gravitational fields). Since all this information—interconnected and communicated by invisible fields—is circulating throughout the universe, we are all to some extent sharing the same information. This is a simple way to understand what is sometimes called the holographic universe.

Sabrina proposed that our challenge as individuals, and as a species, was to learn how to receive and interpret this wealth of circulating information and associated energy. Physics tells us that the info-energy field connections are all there, waiting to be processed and actualized.

I found her logic convincing and compelling.

ON NOT BEING AFRAID TO ASK QUESTIONS WITH HUMILITY: THE BLACK BOX STORY

Soon I would become brave enough to ask the universe a deeply personal and meaningful question. My mind needed courage. Then I remembered my days as a young graduate student when I had been told a visionary story that had remained in the shadows of my mind.

The story was told to me in 1967 by Stephen Krietzman, now a Ph.D., but then a young graduate student in nutritional biochemistry at MIT, in a discussion we were having about the philosophy of science.

I have recalled Stephen's story many times and it has often sparked wonderful conversations. Over the years there were times I told it to students and colleagues; this is the first time I have put my memory of it into print.

STEPHEN'S STORY

A black box fell out of the sky and landed somewhere in the Southwest of the United States. It produced a crater of enormous proportions. The government immediately dispatched investigators to the scene. They fenced off the crater and kept everyone away, save the military and some key scientists and politicians.

Headlines screamed, "What Could the Black Box Be?" Leading researchers were asked to examine it.

First came the physicists, and they did what physicists do: they measured the temperature of the box, bombarded it with subatomic particles, and attempted to reach a conclusion by studying its physical structure.

But try as they might, they could make no sense out of the data they retrieved. The box remained a complete mystery.

Next came the chemists. They did what chemists do. They measured its chemical composition and poured various liquids on it, including strong acids and bases.

But try as they might, they could make no sense of the box, which had a chemical structure unlike any they had ever encountered. They left the site without giving any answer.

Along came the biologists, and they did what biologists do. They searched for signs of biological life, attaching surface electrodes and measuring oxygen consumption, and recording electromagnetic fields, a process similar to recording electrocardiograms and brain waves. But try as they might, they could make no sense of the signals they observed. The box, they said, did not appear to have anything measurable that looked like DNA or cells.

Finally, as a last resort, the officials in charge of the investigation reluctantly allowed a psychologist—we shall call her Dr. Smith—to see the box. She pulled up a chair, sat down, took out a pencil and paper, and then addressed the box in a warm and friendly manner.

She said, "Hi. My name is Dr. Smith, and I would very much like to speak with you."

She paused, and then gently said, "What's your name?"

And the box replied, "Harry."

When I first heard this story, more than thirty-five years ago, it brought a smile to my lips. I remember being clear at the time about the deep lesson this simple story offered: that scientists and laymen alike—all of us—must be open to asking all kinds of questions, even seemingly silly and obvious ones, if we are going to learn and grow.

I believe that people must stay gutsy enough to ask questions that some might judge stupid or frivolous. My personal view is that there are no stupid questions, only uninformed ones—which is the reason we ask questions in the first place: to become informed.

I have tried over the years to apply the black box philosophy to my laboratory research. However, it's one thing to ask questions as a scientist in a laboratory at the university and quite another to do it in the laboratory of one's own mind.

A few years ago, an even more profound lesson from the black box story came to me. I realized that the behavior and communication skills of Dr. Smith were different from those of the other scientists. She asked her questions as psychologists are taught to do—without judgment and in a warm and friendly manner. She put the box "at ease." And she shared something of herself before she asked the box to share in turn. In other words, *she gave first, and then she asked to be given.* And she implicitly invited the box to answer. In fact, clinical psychologists are often taught to use—especially with shy or frightened children—phrases such as "Would you like to tell me your name?"

So, like Dr. Smith, I decided to treat the universe as a black box in a crater, and invite it to give me an answer to my questions. As a clinical psychologist this is what I automatically did when I decided to ask the universe a question. And this may be part of the reason that the universe responded.

6

I Asked the Universe
a Question

G.O.D. IS REVEALED IN THE HIDDEN DETAILS

Sometimes things are better left unsaid.

The exploratory experiment I am about to relate is frankly risky to share—risky not simply because I was open to asking the universe a question, but because the details of the answer I received put pressure on us all to take the idea of actually communicating with a G.O.D. process seriously.

And since it is my personal conviction that scientists are no different from other folks, then I have a responsibility to walk the talk. If scientists take a particular idea seriously, they are obligated to consider exploring it not only in the laboratory but, whenever possible and appropriate, in their personal lives as well. And if I obtain evidence that supports a controversial conclusion, I have the responsibility to share the evidence and my laboratory conclusions in such a way that people can reach their own decision about what the objective findings mean. As a media pundit would say, "Give them the facts without a spin."

I HATED THE WORD "GOD"

Having already learned the lesson of sand paintings and the mathematics of computerized randomized distributions, it was clear to me that there was more to the universe than the science of the day seemed to allow.

I mentioned my interest in a Guiding-Organizing-Designing process to an undergraduate student at Yale who suggested that I might benefit from reading a book about the history of religions. I had no formal training in religion. As a child I had spent some time attending Jewish services but not understanding the Hebrew, and with friends attending Catholic services but not understanding the Latin. Although I traveled extensively, the Bibles in hotel rooms were never opened by me. I did not allow my curiosity to extend to things that I didn't know how to prove, and so I knew less than most of my friends about the origin or evolution of religious thought and practice. But I would learn, and although I forget now the name of my student who recommended religious studies to me, I am beholden to whoever she was.

This same undergraduate recommended that I read Huston Smith's classic *The Religions of Man*. As I always try to follow suggestions—from teachers and frequently from students (who often are my best teachers)—I soon purchased a copy of Dr. Smith's book (no relation to the Dr. Smith of the black box and crater story) and stoically read every word. Reading this book not only transformed my consciousness, but prodded me to do something that I considered to be quite unexpected—at least at the time.

It's difficult to read Smith's book and not walk away with a profound sense of awe and wonderment for the men and women throughout recorded history who have attempted to understand and experience the greater universe in which we live. These so-called prophets, seers, and mystics were friendly to the idea of receiving information from the universe.

If *they* could receive purportedly meaningful information from the Great Spirit, Yahweh, God, Allah, I wondered, should I try, too? Should I try this great "thought experiment"?

I knew that Einstein had taken a ride in his mind with a light beam, and this experience led him on the path to discover the theory of relativity. Inspired by my hero, I had taken a ride in my mind with an electron, and this experience led me on the path to discover the theory of systemic memory described in my coauthored book *The Living Energy Universe.*

I wondered, should I try a personal experiment with the Source? What if I asked the universe a personally meaningful question? Would I receive an answer that was significant? Would it be evidential?

It was approximately 3 A.M. in Guilford, Connecticut. The year was 1986, the season was winter, and it felt as cold as Connecticut winters ever get. My bedroom was as black as I imagined a coal mine would be. If the moon was out I could not see even a sliver of light. It was too dark for shadows and too cold to ponder the mysteries of life. I huddled deeper under the blankets.

Though by day I lived the scheduled life of a tweed-jacketed Yale professor—teaching my courses, directing the Yale Psychophysiology Center, and codirecting the Yale Behavioral Medicine Clinic—at night I would stare at the stars and crave understanding of the organizing process of the All.

What happened that one dark, shivering night, plainly and simply, was that I gave up all pretenses of being critical—meaning judgmental. I, so to speak, surrendered to the All and genuinely asked the universe for help.

Raised as an orthodox agnostic (not meaning one who finds it impossible to know God but rather meaning questioning or wondering about everything, including God), I had never prayed, and this was the first time that I consciously chose to communicate with the universe. What I did that night was an adventure, and I thought of it as taking a ride in my mind.

I remembered the black box and decided to treat the universe in the same way that the fictional psychologist Dr. Smith had treated the black box. I was prepared to be open, honest, and kind.

MY AMUSING CONVERSATION WITH THE UNIVERSE

Completely still in bed, I silently in my head introduced myself to the universe. I shared my name and offered a brief history of who I was.

I then in my mind reviewed with the universe my experiments with sand paintings and computer modeling, and my deep awareness of the need to acknowledge the existence of some sort of Guiding-Organizing-Designing process in the universe (which at the time I called a Grand Organizing Designing process). I explained that I was a systems scientist who appreciated that "the All was in the small." And I confessed that I knew, theoretically, that it was possible to receive information and energy from the universe in the form of communication.

However, I confessed that it was difficult for me to imagine communicating with a universal organizing process because I hated the word "God." I said in my mind: Universe—I have this problem. I hate the word "God." I explained that I was conditioned from the time I was a young boy to think that God was a man with a long white beard and a cane who killed people with germs and bugs (which the Bible renamed as plagues).

I explained that my conditioned response was of "God" not only having threatened, cajoled, and punished people, but even having allowed Jesus to be verbally and physically tortured, and ultimately executed. (As an adult, I had come to understand that there were many other ancient as well as contemporary visions of the Source that seemed to me more mature and enlightened—but that's another story.)

The problem was that I had been well conditioned—like Pavlov's dog salivating to the bell—so that the word "God" evoked a negative emotional reflex. I explained as effectively as I could that these perceptions were preventing me from accepting and exploring a more enlightened vision of universal creative and caring intelligence.

Having now confessed my spiritual problem to the universe, I then asked it for help. I humbly said in my mind, Universe, can you help me? Can you give me another name for God?

Be careful what you ask for . . .

Immediately—and I mean immediately—a name literally popped into my head. The name was not at all what I expected. What I heard was the name "Sam."

Sam! I said to myself, and started to laugh. I knew this was foolish, and tried to reassure myself.

I thought, I finally build up the courage to ask the universe for another name for God. What happens instead is that I end up like a character in a Woody Allen movie.

However, for reasons I did not understand then (but do now), I was moved to get out of bed, to go to my study, and pull out my favorite dictionary, the huge old unabridged version of Webster's second edition. I was compelled to look up the meaning of the word "Sam." It is, of course, short for "Samuel." Sam wasn't listed except as a cross-reference to Samuel. I vaguely recalled that Samuel was one of the books of the Bible, but I had no idea what the name meant, or even if it had a meaning. To say I was shocked is putting it mildly. For a "show me" person, what I was shown that dark night shocked me to my core; I was not even aware of the cold floor or the fact that it had been snowing. The origin of the word "Samuel" was beyond my imagination, my intellectual smarts—beyond belief.

Before I read that dictionary, I had no idea. Moreover, my informal questioning of hundreds of individuals over more than fifteen years since then is that most people have no idea, either; not even many religious Jews know what the name Samuel means or where it comes from.

According to the dictionary, the name Samuel originally comes from the Hebrew Shemuel, which means literally "name of God."

Think about this. I asked the universe for another name for God, and I had received a seemingly innocent, common name—Sam. But it was a word that actually means "name of God."

Being the incorrigible skeptical scientific person that I am, I considered every possible alternative hypothesis I could imagine. And I came up with eleven possible explanations for my hearing the name Sam.

HYPOTHESIS 1—AS A JEWISH BOY I LEARNED THE MEANING AND HEARD THE NAME SAMUEL IN CONNECTION WITH GOD.

Had I heard this information about Sam previously when I went to synagogues for services as a Jewish child?

At age thirteen, I was bar mitzvahed in a determinedly Reform temple. What this meant was that I was required to learn by rote and memorize the sounds of Hebrew. However, I was not taught what the sounds meant. So I never understood Hebrew, nor did I have Bible studies to memorize. I did not know the meaning of Samuel though I was born Jewish.

HYPOTHESIS 2—I LEARNED IT IN CATHOLIC CHURCH.

Had I learned this information when I went to Catholic Church with friends as a teenager? I was raised in a Christian community on Long Island, and when I went with my Italian friends to Catholic services, the services were mostly in Latin. Though I studied Latin for two years in high school, I was never much of a language student. I had no idea what was being said in those church services, and save for *Lux et Veritas* ("Light and Truth," Yale's motto), I know virtually no Latin today. It seemed unlikely that I learned the meaning of the Hebrew word *shemuel* while listening to a Catholic mass in Latin. None of the Christian people I've asked over the years has known the answer to the question "What is the root meaning of the name Sam?"

HYPOTHESIS 3—I LEARNED IT IN *THE RELIGIONS OF MAN.*

Had I learned the meaning of Samuel while reading *The Religions of Man*? Fortunately Smith's book had a good index, so I carefully searched under both Samuel and *shemuel*. Samuel was in the index, referring to that portion of the Bible. However, there was no mention of *shemuel* nor any mention of Samuel being "name of God" in the index or anywhere else in this great book.

HYPOTHESIS 4—IT IS COMMON KNOWLEDGE.

Is the root meaning of the word "Sam" common knowledge? To address this hypothesis initially, I went around the Yale campus and asked students, faculty, and staff, "Do you know the meaning of the word 'Sam'"? Only one student out of hundreds knew, and his father happened to be an Orthodox rabbi. I concluded that the origin and meaning of Sam is not common knowledge.

HYPOTHESIS 5—IT IS A COMMON ASSOCIATION.

Had I guessed the name because it is a common association for God?

In principle, this could be the case. However, when I asked Yale students, faculty, and staff whether they could they give me another name for God, do you think even one person came up with the name Sam? Yale had started as a divinity school; perhaps someone I might have encountered in the early days would have known. But not at that time. "Sam" is definitely not a commonly used substitute for "God."

HYPOTHESIS 6—IT IS A FAVORITE NAME OF MINE.

Did I have a favorite cousin named Sam, a close friend named Sam, a pet named Sam, or a deep affection for Uncle Sam of the United States? None of these were part of my favorites. Nor was I afraid of being drafted.

HYPOTHESIS 7—IT IS A FAVORITE SONG OF MINE.

Having been a professional musician for ten years, I had played and performed many hundreds of songs. I racked my brain trying to find a song that had the name Sam in it. The only song that I could remember that contained the name was "I'm Henry the Eighth, I am." One line said, "And every one was a Henry—Henry—she never had a Willie or a Sam." No, the favorite song hypothesis was not highly probable.

HYPOTHESIS 8—I HAD READ IT IN THE DICTIONARY BEFORE.

Could I have come across the meaning of the word "Sam" while thumbing through the dictionary some forgotten day in the recent past? In theory this was possible, but highly improbable. I do, on occasion, notice a nearby word when looking up something else in the dictionary, but I had never looked up or stopped to notice the meaning of a common name before, or since. This was ruled out, too.

HYPOTHESIS 9—IT WAS JUST A LUCKY GUESS.

Was this just a lucky guess? A fluke? Something that happened by chance alone? The answer is, "As an abstract possibility, of course." However, what is the probability that I would guess this by chance alone? The probability is obviously extraordinarily small.

HYPOTHESIS 10—I READ SOMEONE'S MIND TELEPATHICALLY.

Had I somehow unconsciously read the mind of an Orthodox rabbi or Jewish scholar, living or deceased? Had I somehow read from what Jung termed the mind of the collective unconscious? Had I somehow retrieved the memory from morphologic fields in the quantum vacuum of space, as posited by Rupert Sheldrake in his book *The Presence of the Past*? If the evidence from contemporary parapsychological research is to be taken seriously, then we must entertain some version of a mind-reading hypothesis. However, as far as I have ever been aware, I have little skill for reading people's minds, of this world or the next.

HYPOTHESIS 11—BE CAREFUL WHAT YOU ASK FOR.

Had I received a specific answer to my question from the universe? Had "small" me received a specific communication from the "All," either outside of me or inside of me, or both? I realized that what happened to me on that fateful night was potentially significant not simply because I had received an answer to my strange question, but that the meaning of the specific answer I received could be *verified*.

It is sometimes said that "God is in the details" (though you may have heard it as "the devil is in the details"). Had I not been moved to get up and go pull out my unabridged dictionary, I would never have discovered the potentially evidential nature of what I had received.

I was now faced with a series of questions to answer. Was there more going on here than simple information retrieval? Was the G.O.D. process (aka Sam) listening to one conflicted scientist (me) who was genuinely searching for a new name for "God"? Did the G.O.D. process understand that I required verifiable evidence if I was going to take the information seriously? Did the G.O.D. process appreciate my experimental nature, and did it have an experimental nature too? Did the G.O.D. process play some role in prodding me to get out of bed in the middle of the night and look up the meaning of "Sam" in the dictionary? Did the G.O.D. process recognize that I had an active (and at times silly) sense of humor and would appreciate receiving this particular historic name for "God"?

Needless to say, there is no way from this single self-experiment to determine what the correct explanation is. The obvious solution was for me to conduct more research—to replicate the conditions, first with me as a subject, and then with others.

Do you think I was about to conduct a second experiment? The truth was, I was afraid, and you may wonder what I was afraid of. Was I afraid I was going crazy? Frankly, this thought did cross my mind. But I was well aware that this excuse would be an easy way out.

I was fortunate to know personally some of the best psychiatrists in the country at the time, and I even consulted a few of them about my secret exploratory experiment. I explained the circumstances and what I experienced. The general consensus was that although I was quite stressed by the experience, my exploratory experiment and attempts to understand it did not mean that I required any psychological help.

But I was more afraid that the G.O.D. process might be true. If a G.O.D. process really existed, then I would not only have to change my agnostic approach to life (meaning questioning everything, especially the reality of God) but would also have to face the consternation of my strong atheist colleagues (who represented the majority of the psychology and science faculty).

However, the greatest fear of all was the possibility that I might actually be communicating with some sort of intelligent G.O.D. process, one from which specific information could be obtained that was evidential. This would mean that, in principle, it might be possible for me not only to demonstrate the existence of a universal intelligent Guiding-Organizing-Designing process, but that it might be possible—might even be likely—that I would be able to demonstrate scientifically that the G.O.D. process could personally interact with people on earth—including you.

Such a conclusion would go beyond Einstein's impersonal God that "does not play dice with the universe." This "Experimenting G.O.D." process not only was personal and intelligent, but appeared to have a clever sense of humor!

I would like to state again that the year was 1986 and I was director of the Yale Psychophysiology Center and had research grants from the National Science Foundation and the National Institute of Mental Health. I was codirector of the Yale Behavioral Medicine Clinic and saw psychiatric patients one day a week. I faced an anxiety-ridden dilemma: could I fill out an application to the university's Human Subjects Committee to request permission to conduct experimental research on communicating with "God"?

A WISE DECISION

My decision was simple. I would not touch this question at this time with a ten-foot pole—nor with a hundred-and-ten-foot pole. In fact, I was not going to ask another question of the universe, period. My career would not be put into a tailspin by a Sam or Harry or Professor Huston Smith. Case closed. I even tried drugs for a while.

However, that was then; this is now. And the motto of my laboratory at the University of Arizona is "If it is real, it will be revealed; if it is fake, we'll find the mistake." Time is on my side and the time is right for real information. After all, we're living in the Information Age—it's 2006 and the world is genuinely open to receiving information. And coincidentally Atria Books (my publisher) has published *Tomorrow's God: Our Greatest Spiritual Challenge* by Neale Donald Walsch. If the G.O.D. process or Sam is within reach, then it is up to me and to all of us—including you—to be brave enough, and humble enough, to ask questions and receive information.

It's been said that if we are not part of the solution, then we are part of the problem. This principle may extend systematically, from the Small to the All.

We were walking in paradise,
Never did notice.
Blind in the Buddha land,
Looking for solace.
We had been told of a place,
Far beyond this vale of tears.
We could never have guessed,
We were already blessed.
There we were, where we are,
In the garden.

FROM JAMES TAYLOR, "UP ER MEI,"
ON THE ALBUM *HOURGLASS*

7

I Asked the Universe More Questions

THE ANSWER SPARKLES LIKE A DIAMOND

I have quietly and privately conducted numerous exploratory experiments on this topic in the privacy of my home, on mountaintops, in hotel rooms at scientific meetings, at spas and resorts, but I have yet to initiate a formal university research program on the topic of "asking the universe questions." This will require human consent forms and Internal Review Board endorsement. But the time will soon come.

It was almost a decade after my original Sam conversation at Yale when I felt the need to initiate asking the universe some questions again. And to my astonishment, again I received answers. (Incidentally, I sometimes refer to the universe as "universe," sometimes as "Samuel," and sometimes affectionately as "Sam," which is non-gender-specific, since it could mean "Samantha." It doesn't seem to make a difference. Answers come to me so long as I am genuine, I surrender, I am open, I am humble and nonjudgmental.)

There was always the possibility that my experience with hearing the "Sam" was a onetime occurrence, and therefore was coincidental, as in unrepeatable. That turned out not to be the case. Recently it has become essential to me to reveal the truth. But that doesn't mean it's my responsibility to share every shred of evidence I have received over the years. Between the constraints of space and, sometimes, in respect for people's privacy, I include in these pages only what I deem to be appropriate. As Professor William James put it, there is wisdom in knowing what to overlook.

Late one afternoon while writing this book, I reminded myself that shortly after Christopher Robinson and I had completed the "Ten Days in Arizona" experiment, he confessed to me that he personally saw the greater purpose of our experiment as a "search for God." I wondered at the time if Sam had sent Christopher to me in answer to my continued quest for tangible evidence of G.O.D.

THE MEANING OF THE DIAMOND

Alongside my OfficeJet printer in my home office as I write these words sits a special piece of Tiffany cut glass that is the size of an apple but shaped like a diamond. It's one of those paperweights that is impossible to miss and brings "ah's" from visitors. When the OfficeJet prints, my Tiffany "diamond" wiggles and its facets reflect light in various directions. While waiting for words to come to mind I sometimes gaze at this fine glass object, and often, when the light comes in at sharp angles, I see many colors in each of the facets. And often the rainbow of light within my paperweight takes my thoughts back to that astounding conversation I had with Sam and the remarkable confirmation that occurred the next morning. Now I need only look at my Tiffany glass diamond to get a feeling of being in touch with the universe.

I purchased my Tiffany glass diamond a few months after I received a surprising and special answer to my unanticipated question of the universe. It was April 2001 in Albuquerque, New Mexico, and I had been invited to give two addresses at a conference on the science of consciousness. One of my talks was to be on *The Living Energy Universe;* the other would be on *The Afterlife Experiments.*

When I finished the *Living Energy Universe* talk, to an enthusiastic standing-room-only crowd, a rush of people wanted my personal attention; they had come to the podium to probe more deeply into the topics of my talks. A tall blond woman, whose sharp bone structure was matched in beauty by her intense and vital energy, raised her voice to be heard. "Do you really believe what you've told us? Are you saying there's really a living energy universe? And do you really believe in survival of consciousness after death?"

I told the woman that as a scientist, what I would be willing to say at that point in time was that "when all of the data are considered, the simplest—and most parsimonious—explanation that accounts for the largest amount of the data is the one I gave in my talk." Then I told her in an equally clear and loud voice that I did agree with survival of consciousness after death as a viable hypothesis.

She was not yet satisfied, and pressed on. "But do you *believe?* What do you personally feel?"

Believe? Feel? Scientists are not supposed to believe or feel our way to conclusions. What science is supposed to do is question, investigate, and evaluate alternative hypotheses. Scientists are taught to base conclusions on facts and research but not on feelings or belief systems, which should not be brought into the laboratory.

I suggested to her that we put off the conversation to another time, so I could answer other questions. But I was troubled by her challenge and the fact that I had set her curiosity aside.

AN ANSWER FROM SAM

I realized that I had been haunted by the mystery woman's question of whether anyone can know scientifically that information and energy truly exist as invisible fields in the "vacuum" of space, as well as the "vacuum" that comprises matter. Quantum field physics tells us, in no uncertain terms, that matter is mostly empty space that is organized by invisible fields.

Returning from the speech I carried back with a responsible weight two heavy-duty questions. Does science know that informational and energetic fields play a fundamental role in shaping all things, great and small? And is a creative, intelligently organizing field process involved in the manifestation of all things? These were questions I had been pondering the previous night.

So, in the privacy of my hotel room, I decided to raise these questions with the universe. In a state of complete surrender and openness, prepared to hear whatever popped into my head, I asked Sam one question.

"How do I know, and how can I prove, that external organizing forces, mediated by invisible fields, play a role in the creation of all things, including physical matter?" In everyday language this means "How do I know that you (G.O.D.) are real?"

Scientists don't typically seek proof but depend on collected data and then determine whether the data are consistent or inconsistent with the hypotheses. But I wanted more. I wanted to know what to believe, and I wanted to know how to reach a decision.

What I heard from the universe, as before, came as a stunning surprise. In fact, it shook me to my core. I heard, "Remember the diamond." Three simple words—but a serious answer that was another mystery to be solved.

For the record, I didn't own a diamond ring and had no intention at that time of buying one, not for myself, not for any lady friend. The idea of diamonds could not have been further from my conscious mind when I asked the universe the question. So I repeated the words, a reply that was a question. "Remember the diamond?"

As I spoke those words, three sets of visionary images flashed through my mind. The first was the fact that diamonds do not simply appear out of nowhere. The creation of one of these precious stones requires (1) extensive pressure (2) from the outside (3) for long periods of time (4) to collapse carbon into the supersolid state called a diamond. In my mind I experienced the carbon being organized through the external forces upon it. I felt myself, so to speak, becoming a diamond in the rough.

Take-Home-Message One: Patterns of outside force are required to give birth to the diamond in the first place.

The second image that came to me was of a diamond cutter planning to sculpt the raw diamond into a sparkling gem. The diamond's potential to reflect light deserves the term "awesome." However, a visionary and skilled diamond cutter takes a rough diamond and reveals its inherent potential for beauty. Diamonds stand alone in their combined magnificence and resilience. Mentally I experienced three things simultaneously—the rough diamond, from its own perspective, being turned into something spectacular . . . the diamond cutter, from his perspective, seeing his vision manifest through his hands . . . and an invisible outside intelligence smiling at this remarkable creative process (from Sam's perspective?).

Take-Home-Message Two: Humans, using external force, continue the organizing process through deliberate planning and design. Moreover, this planning and design process is potentially co-creative with an invisible intelligence.

The third image that came to me was of a scientific satellite, the Cosmic Background Explorer (COBE) telescope, floating in space, taking pictures of the background radiation—infrared patterns of light at very low temperature, present everywhere in the universe. The COBE telescope was a stupendous human achievement of innovation, planning, design, and implementation.

What the COBE telescope revealed was that the background radiation of the universe is not random. The background radiation, purportedly 12 billion or more years old, was supposedly created shortly after the birth of the universe—the event that has been playfully termed the Big Bang. The background radiation is actually composed of photons—massless particles of light that are invisible to the naked eye but are detectable and ultimately displayable by means of modern technology. These ancient photons were clearly organized, expressing a pattern that is more than 12 billion years old (the exact age is debated). Do you know what the detector is made of that reveals these 12-billion-year-old organized photons? The answer is, the diamond.

In my mind I literally experienced myself in space. I was both inside the COBE telescope, inside the diamond detectors, and outside the telescope as well, witnessing the map of the background radiation present throughout the universe.

Take-Home-Message Three: The diamond reveals patterns of organized creation—in the "vacuum of space"—that go back to the beginning of time as we know it.

I trust you can sense the depth of awe and wonder I experienced when those three sets of images flashed through my mind: first the formation of the rough diamond, then the diamond being carved, and finally the use of the diamond to reveal the invisible order of the background radiation some 12 or more billion years old. The combined process literally took my breath away.

I said to myself, I must remember the diamond. I was struck by the thought that the diamond could serve as a symbol for an intelligent Guiding-Organizing-Designing process that potentially existed since the beginning of time. Let me not forget to include here what I later learned—that ancient religions viewed the diamond as a fundamental expression of "God."

I decided that night that someday I would purchase a real diamond to remind me of my special symbol as well as this momentous experience.

I reasoned: if I can believe in the reality of diamonds, and what they reveal, then I can believe in the existence of an organizing universe that contains intelligent, dynamic, evolving fields of energy. And in the process, I realized that I had an answer to the blond woman's question. Yes, even as a scientist, the time had come when I could believe and say so without apology.

The next morning, on the way to another of the conference presentations, I encountered the same woman, and once again she challenged me: "Well, Dr. Schwartz, do you believe?" But the answer I gave was not one I had planned or expected. My answer was one I would never have predicted. I looked furtively around, like a character on *Law and Order,* and determined that we could not be overheard. I was about to give her an answer that I had received to a question I had asked of Sam the previous evening.

THE UNBELIEVABLE IDENTITY OF THE MYSTERY WOMAN

Though I was not prepared to meet her so soon (if ever), I was ready with an answer to her question.

Needless to say, I was not about to tell her *how* the answer came to me, nor would I share the phrase "Remember the diamond" and how I was given it as a symbol for my answer.

I again explained to her that scientists are not supposed to believe. We are supposed to experiment, to investigate. However, I did admit that there must come a point in time when even a scientist has enough logic, data, and direct personal experience—the holy trinity of learning—to be able to reach a firm conclusion.

Knowledge is like a three-legged stool—one leg is logic and reasoning, one leg is data and experimentation, and one leg is intuition and experience. When all three legs are present, science can stand on a knowledge stool that is well balanced and firm.

She listened carefully to my explanation that from this vantage point, I could honestly say that I believed in a living energy universe (from the title of my book that describes how energy and information express the fundamental properties of life). I was now prepared to definitively conclude that some version of the theory was true—including survival of consciousness after death and a G.O.D. process.

She smiled. And then she told me that her name was Barbara, that she lived in New York with her husband, and that he had come with her to the conference.

While we were talking, the woman's husband joined us. She introduced him to me as Bruce Winston and I said, "Hello." The Winston name did not sparkle in my mind—not yet.

She said, "You're aware that Bruce is very well known, but he doesn't like to talk about it."

I confessed that I did not recognize the name.

Barbara said, "Bruce is Harry Winston's son."

I admitted that I was still in the dark.

Barbara looked at me with amazement. She said, "You don't know that name? Harry Winston was one of the largest and most successful diamond dealers in the world. In fact, he was the man who donated the Hope diamond to the Smithsonian."

Diamonds, again!

In the back of my mind was the memory that Sam had said I should "Remember the diamond." And also that I had just decided to purchase a small diamond ring someday to wear as a reminder. Now I had met the son of one of the most successful diamond designers and dealers in the world. At that moment I didn't classify these events as a coincidence. I felt as if I had been assaulted by the events and the experience—of Sam, diamond, and Winston. And, as well, by my open expression of my belief.

I did not say anything immediately to the Winstons about the remarkable synchronicity of all this. However, after Bruce left, I decided that maybe I should confess the "Remember the diamond" secret to Barbara. I explained how her persistent questioning the day before had challenged me. I described the unplanned conversation with the universe that night (but did not tell her about Sam). And I shared the three images that came to me when I heard the phrase "Remember the diamond." Barbara was touched. She told me that when I was ready to purchase a diamond ring, I should purchase it through Bruce.

But as you already know, I subsequently did purchase a diamond—albeit a cut-glass one at Tiffany's rather than a pricey, pure white cut diamond in a pinky ring. My "hope" diamond paperweight reminds me not only to "Remember the diamond." It reminds me of the larger meaning and context preceding and following my surprise conversation with the universe: that the evidence for the existence of an experimenting G.O.D. process exists everywhere—not only in sand paintings and clothing, but in diamonds and the intertwining of people's lives.

REMEMBERING TO BE CAUTIOUS AND OPEN TO ALTERNATIVE EXPLANATIONS

For the record, we cannot conclude from this incident that the phrase "Remember the diamond" necessarily came from Sam, aka the universe.

The superskeptic could speculate that somehow I had learned that the Winstons were staying at the hotel before I asked the universe my question (and furthermore that I actually knew who Harry Winston was but had forgotten this information). He could speculate that it was just a fluke that the woman who asked the question happened to be connected to diamonds. The more adventurous skeptic could speculate that I had somehow read Barbara's mind telepathically and unconsciously learned of her association with diamonds. The truly courageous skeptic could even speculate that maybe I received the information from the deceased Harry Winston, who knew that his son and daughter-in-law were at the conference, and had witnessed Barbara asking me her important question!

At that time, all I could say with certainty was that I had innocently posed a question in my mind to Sam and that I immediately heard or experienced three words that caused me to have deeply meaningful images of diamonds—from the earth, through the gemstone designer, to outer space.

A very spiritual friend of mine had e-mailed me a quote that she thought I would enjoy: "There are no coincidences. They are miracles for which God doesn't want to take credit."

Consider the combination of (1) receiving surprising and meaningful information about diamonds (2) in response to a fundamental question from an unknown woman (3) who subsequently turned out to be the wife of one of the heirs to the Harry Winston diamond heritage. This pattern certainly leads one to wonder, Where is the information coming from that Dr. Gary Schwartz is receiving?

IS EVERYTHING, METAPHORICALLY, A DIAMOND?

When we see a diamond, are we being shown, metaphorically, the face of the G.O.D. process?

Contemporary science suggests that when we look at anything, we are potentially seeing examples of an infinite number of faces of what I have named the Guiding-Organizing-Designing process. This creative unfolding process, this seeing beyond the mere physical presence of the object, is flexible, complex, and sophisticated. And so for me, a diamond has become a master metaphor—a material symbol of the existence of universal intelligent organization expressed as exquisite beauty and sensitivity, in nature and in the cosmos.

It is true that nonrandom natural forces create the diamond in the rough. But it is humans who first saw the potential beauty of the diamond that is cut and shaped and perhaps set as a Winston might perceive it. It is a human who put a pattern of sparkles into a raw diamond by shaping it. And it is a human scientist who can use the diamond to potentially reveal the living history of the very creation and organization of the universe.

Is this potential ultimately a reflection of the mind of the G.O.D. process, expressing itself through the mind of human beings? Did the G.O.D. process create the diamond partly as proof of his existence? So that, in contemplating it, people like me would find ourselves contemplating the G.O.D. process? Perhaps the same is also true of a tree, a flower, a horse, an amoeba? Are humans as a species diamonds in the rough, readying to be transformed into gems of indescribable brilliance? Is the human race beginning to awaken to a divine garden that is already here? Sir Isaac Newton supposedly said, "I don't know what 'synchronicity' means—but it's the third time I've heard it this week."

Paraphrasing James Taylor, I can honestly say that the following applies to me. And maybe to you, too?

I could never have guessed,
I was already blessed.
There I was, where we are,
In the garden.

A garden of flowering diamonds . . .

There are only two ways to live your life. One is as though nothing is a miracle. The other is as though everything is a miracle.

ALBERT EINSTEIN

8

Interesting and Amusing Theories

A PHYSICIST CONCLUDES, "IT'S ABSOLUTELY WORTH THINKING ABOUT."

Imagine that you are interested in the fundamental question "Is there evidence for an Experimenting G.O.D. process—an intelligent Guiding-Organizing-Designing process—in the universe?" and the related question "Can we communicate with it?"

You have just received another seemingly impossible answer from Sam. You are convinced that a miracle has occurred. As a scientist you have finally received major funding for your visionary and daring "Experimenting G.O.D. Field" project. The donors who will fund the project accept the fact that fields exist everywhere, including in the "vacuum of space." These backers further accept the fact that fields are forces that carry information and serve the purpose of communication in everything—from individual atoms to superclusters of galaxies, and everything in between (including us).

They further agree with you that a "field," as physicists use the term, has the potential to integrate the primary universal processes attributed to "God"—purpose, intelligence, and power. And they agree that if there is a G.O.D. process in nature and the universe as a whole, then that G.O.D. process must somehow involve invisible fields—since everything that is material emits fields and responds to fields.

Your scientific goal is to integrate these two major ideas and conduct systematic research that addresses the question "Is there a measurable G.O.D.-field process in the universe?" Your task is to determine whether or not some kind of purposeful, creative, and caring field process is playing a guiding role in the design and evolution of the cosmos—an intelligent trial-and-error process we humans call "experimenting."

What kind of evidence will you decide to gather? What kind of experiments will you run? And as you gather the evidence, how will you know if your Experimenting G.O.D.-field process interpretation of the evidence is correct?

Even though we have already considered evidence that points inexorably to the existence of some kind of Experimenting G.O.D.-field process in the universe as a whole, it's instructive to step back and take a deep breath. A few more questions of a general nature are necessary. How do we search for evidence scientifically? And how will we interpret what we find? What criteria will the team choose to use to conclude that a given explanation of evidence is reasonable and credible?

To understand how even the most conservative scientific investigators sometimes arrive at bold and far-reaching conclusions, it's meaningful to consider how responsible science goes about the process of entertaining, rejecting, and accepting hypotheses, especially extraordinary ones.

Partly for entertainment, and partly for clarity, I'd like for a moment to explore with you the science of hoofbeats, horses, zebras, and the Experimenting G.O.D.-field process.

ZEBRAS IN CENTRAL PARK?—A LESSON REGARDING EVIDENCE VERSUS EXPLANATIONS

Emergency room doctors have a saying that has apparently been passed down through generations: "When you hear hoofbeats, don't think zebras." Physicians, especially ER doctors—who often must play God to save their patients—appreciate the deep wisdom of these words. But what do they mean?

Imagine that you are in New York City, near Central Park, and suddenly you hear hoofbeats. What's the most probable explanation for this sound? The answer, of course, is horses—not zebras. Horse-drawn carriages are a possibility, as are mounted police; both are part of the city scene. Except at the zoo, or if a circus is in town, you are not likely to see zebras in Central Park. Now, if you hear hoofbeats but don't see any horses, you would then consider other less probable but still likely possible explanations, such as ponies, donkeys, or mules. However, if you can't find any ponies, donkeys, or mules, then you might consider the possibility that what you actually heard was a zebra—perhaps one that escaped from the Central Park zoo or the circus.

Detectives following in the footsteps of Sherlock Holmes appreciate the importance of considering all possible explanations for a given observation or phenomenon, and then systematically ruling them in or out through the careful scrutiny of the clues and evidence. In other words, a conservative approach to searching for evidence, as well as drawing conclusions from the evidence, is to consider the history of what is known about a phenomenon in a given context, and begin your search with the most probable explanation.

Imagine that you are an emergency room physician facing a life-or-death situation, and you must make your decisions quickly. What rules should you follow to optimize saving a patient's life? The answer is simple. Pick the most probable explanation or cause for the symptom based upon years of accumulated clinical and experimental evidence, and check to see if this specific cause is the explanation. If the most probable explanation turns out not to be correct, you will rule out that explanation and move on to the next most probable one.

The hoofbeat lesson works in science, too. It's prudent to examine the most probable and accepted explanations, and only after ruling them out, consider the less probable and more controversial explanations of the phenomenon.

FROM ZEBRAS TO THE EXPERIMENTING G.O.D.

Consider the Experimenting G.O.D. process question. Atheists as well as conservative scientists might say, "When you see evidence for the existence and evolution of order in the universe, don't think 'God.'" Another way they might say this is, "When you see evidence for the existence and evolution of order in the universe, don't think universal intelligent design."

However, there's an important distinction between atheists and conservative scientists in how they qualify this phrase. Atheists simply assume that there is no "God"—that universal intelligent design in the universe is impossible—hence they dismiss the G.O.D. process possibility out of hand. For them, there are no zebras; only horses exist. For atheists the statement would read: "When you see evidence for the existence and evolution of order in the universe, think *chance. Period.*"

Conservative scientists take a different tack. They propose that the chance explanation is the most probable, and therefore the existing evidence must rule out the chance explanation before they will have much interest in considering other possible explanations. For the conservative scientist, it's possible that the hoofbeat sounds are being made by a zebra, but in order to consider zebras, they must first rule out horses. Hence, for conservative scientists the statement should read, "When you see evidence for the existence and evolution of order in the universe, think chance *first.*"

As a conservative scientist, I agreed with the approach "Think chance first" when I asked the question "How do we explain the evidence of order and evolving order in the universe?" In fact, I originally did what most skeptical scientists do: I presumed that the chance explanation was the most plausible and probable explanation.

One of the greatest surprises revealed to me was that the chance explanation turns out to be neither plausible nor probable. What I discovered, quite unexpectedly, is that this explanation simply doesn't work.

Recall that when you actually do experiments to test the chance hypothesis—for example, to see whether sand paintings will create themselves by chance, whether computers generating random numbers will create shapes other then a bell-shaped curve (described in Part Two)—they always fail to produce an evolving order. The experiments fail to produce an evolving order 100 percent of the time. All things being equal, in the absence of a Guiding-Organizing-Designing process, the experiments confirm that most things mix and become less ordered, not more ordered (in physics this is called the Second Law of Thermodynamics).

To return to our metaphor, for centuries science has assumed that the chance explanation was a "horse" but the horse surprisingly turns out to be a Trojan horse. Chance is not what we have assumed it to be.

The history of science reminds us that we should always strive to be cautious, proceed with humility, and maintain an open mind. Hence, when I offer extraordinary evidence, inexorable logic, and highly supportive personal exploratory experiments in this book, I do so fully aware of the possibility that the seeming stampede of hoofbeats we seem to be hearing may eventually turn out to be from something other than a herd of zebras. In principle, it's theoretically possible that the conclusions in this book may turn out to be merely amusing or, better, not just amusing but actually true. However, we should always be prepared for the possibility that new evidence may come along in the future that indicates that it was a horse after all, or that it was neither a horse nor a zebra, but was a camel, or even a unicorn, instead.

EMOTION AND THE G.O.D. QUESTION

The question of the existence of an Experimenting G.O.D., a creative Guiding-Organizing-Designing process, may be the most emotional of all questions humans think about. Some people experience joy, wonder, and peace when they ponder God. Others experience anger, fear, and disgust. Some are fearful of the idea that an invisible power may be guiding their lives; others *depend* on the idea; still others label the God idea as superstitious if not outright foolish. Finally, there are those who literally hate the idea of a G.O.D.: some conservative scientists who will remain nameless have told me of their strong distaste for the existence of a G.O.D. process.

Depending upon your current religious or philosophical beliefs about "God"—complete acceptance, openness to questioning, or total rejection—you may be experiencing feelings ranging from inspired to infuriated as you ponder the evidence.

If you already believe in "God," you may be finding the evidence comforting and validating. If you are currently agnostic and don't know what you believe, you may be finding the evidence helpful if not transforming. If you are an assured atheist, you may be finding the evidence challenging and even, I'm sad to say, potentially threatening. This is unavoidable—it comes with the territory of this research. In fact, if you consciously (or unconsciously) dislike the Experimenting G.O.D. explanation, you may be engaging in selective use of logic accompanied by actual blocks in your understanding.

A dear friend who will remain nameless once said to me, "When I dislike something, it stops me from understanding it." Another dear friend who will also remain nameless said to me in one breath, "Gary, everyone knows that clothing does not fold by itself," and then, in the next breath, he said, "But if we wait two billions years or so, maybe the clothing will fold by itself." A third friend was annoyed that I would even search for evidence, saying that her belief depended on an absence of evidence. "Finding evidence would be like what happened when a man walked on the moon; I never looked at the moon romantically again."

Your assumptions are never based solely on evidence or reasoning; they are based on emotion and wishing as well.

"I FIND YOUR THEORY MOST AMUSING"

In 1982, while I was preparing to give my presidential address to the Health Psychology Division of the American Psychological Association, I wondered whether to share my burgeoning interest in how quantum field physics and relativity theory could be applied to health psychology and the psychophysiology of healing; I knew that even to suggest the application of quantum physics to proving anything about health or healing would be highly controversial at the time.

I referenced a book I had read years earlier; *The Cosmic Code,* a book I mentioned in Chapter 4, pursues the history of quantum physics. I had been taken with the description of how Einstein had developed his field theory of relativity. The author, Heinz Pagels, was, as pointed out earlier, a distinguished physicist and professor at Rockefeller University.

Following the model provided by Pagels, I put myself in Einstein's shoes, so to speak, and extended Einstein's logic to physiology, personality, and psychology. This resulted in what I termed the general relativity/general relaxation hypothesis. But I was not sure about my logic and my predictions and wanted to test them.

The details of the controversial theory and the significant evidence I collected that supported my theory were later published in a 1987 book titled *The Emergence of Personality.* What is most important is what happened when I decided to call Dr. Pagels, who was at that time still alive, to ask him if he agreed with my logical extension of what he had written. I wanted his verification and so I contacted him to discuss the theory.

He spent an hour on the phone with me. In the first thirty minutes, I explained to him that this speech was an important one to me and to the American Psychological Association. I needed his advice as to whether I should share this original theory with my colleagues since I knew that some of them would be highly critical of me for doing so. I then went through the theory in detail, step by step.

When I was finished, Pagels merely said, "I find your theory most amusing."

There I was asking a distinguished physicist I held in high esteem for his scientific advice and personal wisdom about whether I should share my thesis in a major address. His response of "most amusing" was both shocking and disheartening.

But Pagels continued, to be sure I understood. He said, "Gary, you don't know what I mean by the word 'amusing.'"

He explained that as a graduate student at Princeton University, studying with the Nobel Prize–winning physicist Eugene Wigner, he learned that Wigner was fond of telling his students that there were two kinds of theories in the world, one of which he described as "interesting" and the other "amusing." Wigner would explain with a twinkle in his eye that "interesting theories, though often true, are typically not worth remembering, whereas amusing theories, though often not true, are *absolutely worth thinking about.*"

Hearing that story, I breathed a sigh of relief. This great man had used the phrase "most amusing" as a compliment! So I decided to share with my colleagues my intention to amuse them by applying quantum physics and relativity theory to health psychology. As I had anticipated, some of my colleagues were excited and inspired; others were quite angry and dismissive.

So, dear readers, using the same terminology, this book is meant to be amusing in Pagels's definition of the word. The evidence, logic, personal experiences, and implications, I hope you'll agree, are absolutely worth thinking about.

Can the minds that created multigigahertz portable computers, crystal clear global satellite cell phones, ultrahigh resolution pocket digital cameras, and wireless personal digital assistants connected to the World Wide Web have the potential to discover the origin and meaning of our own intelligence? Is it possible to uncover the apparent intelligence and power of the cosmos as a whole?

Part Four

THE GREAT G.O.D. DEBATE

What is arguably the greatest debate is whether the existence and evolution of order in the universe—including our personal experiences of coincidence—can be explained solely in terms of chance. Or does it require the presence of creative intelligent design and an Experimenting Guiding-Organizing-Designing process?

And are physical, biological, and social evolution all expressions of intelligent evolution?

Unthinking respect for authority is the greatest enemy of truth.

ALBERT EINSTEIN

9

Chance Versus Intelligent Design—Which Is It?

WHAT IS THE GREAT G.O.D. DEBATE, AND HOW CAN IT BE RESOLVED?

Okay—up to now I've presented exemplary evidence as well as compelling theory pointing to the conclusion that some sort of a Guiding-Organizing-Designing process is playing a fundamental role in the origin and evolution of the universe—as well as in you and me. At this point I want to step back and consider the question of chance versus intelligent design in more detail. I ask you to follow this in the spirit of the distinguished neuroscientist Warren McCulloch, who said with great fondness, "Do not bite my finger, look where I am pointing."

There are basically two schools of thought—what scientists label as two foundational hypotheses—about the origin of the universe. One can be called the "chance universe"; the other, the "intelligent design universe." Though there are many variations and possible combinations of these two broad classes of theories, the complexities boil down to two simple hypotheses.

The chance universe is the current interpretation generally accepted by mainstream science in the fields of statistics, quantum physics, and evolutionary biology. It suggests that the growth and evolution of all things, from the subatomic to the galactic—and everything in between—reflects a version of natural selection and survival of the fittest as elegantly proposed by Charles Darwin. At its core, the idea of survival of the fittest is presumed to be possible due to the existence of random mutation and chance in biological systems, driven by physical laws, as they interact with the environment.

Highly regarded twentieth-century scientists—ranging from the late Richard Feynman and Carl Sagan in physics and astrophysics to Stephen Dawkins and the late Stephen J. Gould in evolution and biology—are often associated with some version of this randomness interpretation of the origin of order, life, and evolution in the universe.

It is important to pay close attention to the wording here—we are speaking about the origin of order, life, and evolution in the universe. Generally speaking, mainstream scientists tend to ignore the "origins" question because they cannot address it. Explaining the evolution of biological life in terms of existing laws of physics, chemistry, and biochemistry ignores the fundamental question: what explains the origin and organization of the laws that are presumed to be the building blocks of higher-order biological systems, including the appearance and evolution of intelligence in life? If the explanation for the origin and organization of natural laws is not chance and randomness, then what is it?

Other distinguished twentieth-century scientists—ranging from the late Albert Einstein and David Bohm in physics to Rupert Sheldrake and the late Willis Harman in evolution and biology—are often associated with some version of an intelligent design interpretation. A more apt term, one that is consistent with the evidence from evolution as well as from statistics, is "creative intelligent design." This intelligent trial and error or "experimenting intelligence" integrates intelligence as expressed in both art and science.

The proponents of the chance universe hypothesis cite veritable mountains of data that *seem* to be consistent with the existence of randomness in the evolution of the universe. I emphasize the word "seem" because to accept the chance hypothesis requires that one assume that if something appears random—like the sequence of numbers in pi—then the order must be random (a very shaky assumption). It also requires the assumption that the overabundance of order seen in the universe (for example, spiraling galaxies) can be explained as random combinations of laws and processes—yet this is inconsistent with everyday experience revealed through sand paintings or watchmaking.

The most vocal advocates claim—with complete conviction—that randomness essentially rules the universe. They presume that the universe began in a state of true chaos (disorder), and that it is basically running down (again, this is the Second Law of Thermodynamics). The seeming exception to this overall picture of increasing entropy and disorder in the universe is the existence of small pockets or areas of evolving order—such as witnessed on the planet Earth.

To the chance advocates, if such a thing as "God" exists, there would be no question that such a God plays dice with the universe. In fact, according to some interpretations of probability theory, you can literally bet your life that the chance universe hypothesis is true.

However, before we can address the playing dice explanation, we must consider the distinction between observation and interpretation, between the facts and our stories about them. Remember we *observe* that pi produces a precise sequence of numbers that defies our current understanding, and yet this sequence is replicated 100 percent of the time. What is our *interpretation* of these observations? Randomness? Complex order? Expression of intelligent design? How do we decide among these potential interpretations?

HOW LOOKING AT THE SUN HELPS US UNDERSTAND THE FUNDAMENTAL DISTINCTION BETWEEN OBSERVATION AND INTERPRETATION

Just because an *observation* is replicable does not mean that our *interpretation* of the observation is correct. And this deserves repeating because if there is one lesson we should all learn from the history of science, it is this: *The fact that an observation is replicable does not mean that our interpretation of the observation is correct.*

Unfortunately, even the best-trained scientists sometimes forget this core distinction. One of my favorite examples of the "interpretation is not the observation" principle involves commonplace observations of the sun that lead us to say the sun rises in the east and sets in the west.

For almost two decades I have lived in a beautiful desert where the sun shines almost every day. I have personally witnessed, and can easily document over three hundred times in a given year, the sun's apparent rising and setting behavior. Just like flipping a coin or throwing dice, no expensive and complicated scientific devices are needed to make observations of the sun's trustworthy behavior.

These completely replicable solar observations long ago led humans to the seemingly commonsense interpretation that the sun revolves around the earth. So the distinction between observation and interpretation is critical. Early humans witnessed the sun's rising here and setting there, and inferred—that is, formulated the conclusion—that the sun "revolves" around us.

History teaches that for many thousands of years, this is exactly what humankind did. Early man confused the distinction between observation and interpretation and accepted the interpretation that the sun revolves around the earth as if it were an observed fact. We still retain that notion when we talk about the sun rising and setting, as if it were moving and the earth were still.

But the interpretation is not the observation. The conclusion is not the data. The map is not the territory. The explanation is not the evidence.

We now understand and accept that the most obvious interpretation of the sun's rising and setting behavior turns out have an alternative explanation. But it's well worth remembering that this interpretation not only challenged common sense, it challenged long-held religious beliefs based on the commonsense interpretation, as well.

We now understand that because the earth revolves on its axis, and because the earth revolves around the sun, it therefore appears to us on the earth as if the sun is revolving around the earth, when in fact something quite different is happening.

It's also worth considering that any "alternative" interpretation was once extremely dangerous for the scientists who proposed it, because the establishment—in this case, the orthodox church—was committed to a different interpretation. In fact, in the course of human history, scientists have sometimes been silenced, tortured, and even executed for proposing interpretations that conflicted with accepted foundational interpretations of "the establishment."

Informed people have come to know what the "alternative" interpretation is of the sun and earth. The names Copernicus and Galileo are connected to one of the greatest paradigm shifts in the history of science, a shift often referred to as the Copernican Revolution. And just for the record I will state what is obvious to most twenty-first-century adults, something contemporary science has convinced us to take for granted: that there is a logical, plausible, and convincing alternative explanation for our earthbound observations of the sun.

Note that the basic observations of the sun's behavior—seen by our eyes when on earth—have not changed. What *has* changed is our intellectual interpretation of our earthbound, and hence earth-limited, observations of the sun. Unfortunately, when specific interpretations of observation become accepted dogma, the human mind typically closes shut, and creativity atrophies. Einstein put it this way: "Unthinking respect for authority is the greatest enemy of truth."

A NEW, ALTERNATIVE INTERPRETATION OF RANDOMNESS

What you are about to read is an alternative interpretation of coin flips, dice, and so-called randomness that is parallel to the alternative interpretation of the sun revolving around the earth.

In a fundamental sense, our long-held interpretation of the observations of coin flips—as chance—is as primitive as man's early interpretation of the observations of the sun. I'm volunteering to lead the charge and to put new interpretations onto old observations.

The alternative interpretation revealed in the next chapter resolves the apparent paradox between a young child seemingly throwing random dice and the Dream Detective (in Part One) seeing a clearly non-random future.

Can a simple change of concept that takes place in the mind transform our perception of virtually everything? The history of science says yes.

My religion consists of a humble admiration of the illimitable superior spirit who reveals himself in the slight details we are able to perceive with our frail and feeble minds. That deeply emotional conviction of the presence of a superior reasoning power, which is revealed in the incomprehensible universe, forms my idea of God.

ALBERT EINSTEIN

10

Can G.O.D. Play Dice with the Universe?

EINSTEIN WAS RIGHT: THE ANSWER IS "NO"

You now have the background knowledge for understanding a new interpretation of randomness, and for understanding how this interpretation is consistent with the G.O.D. process explanation. In this chapter I follow the logical progression by showing how G.O.D. could have designed into the very fabric of the universe the capacity for apparent randomness. And I will show how this apparent randomness may bring relative freedom, discovery, and creativity.

When Einstein said, "God does not play dice with the universe," he expressed a "deeply emotional conviction." However, there were "slight details" that revealed God's presence to Einstein. The logic to him was inexorable. But the rest of us were not able to follow the logic with what might be called our "frail and feeble minds."

I want to consider two such details that empower us to correct the long-standing mistake in our conception of how randomness works, as well as to further explain why randomness cannot be used to explain the existence of order and evolution in the universe. The answer takes us to the hypothesis of intelligent evolution.

TWO ESSENTIAL CONDITIONS FOR THE EXISTENCE OF APPARENT RANDOMNESS

Every student of statistics is taught that there are two keys—two fundamental conditions—that must be present in order to create a normal, or bell-shaped, curve of numbers. These two conditions are so well accepted that they are taken to be assumptions—we simply accept that they are true.

Using coin flips as an example, the two keys that statistics teaches are as follows:

1. As discussed earlier, each coin flip must be *independent* of every other coin flip. Each time the coin is flipped, the flip must be independent of all the flips that have happened previously. The fact that the last flip was heads must not influence the way you produce the next flip. Complete independence is required to obtain a normal distribution of coin flips. If the events are *not* independent, every possible combination of coin flips will *not* have the "freedom" to potentially occur, and this freedom is an established requirement in statistics.

2. Both the coin and the coin flipper must be *unchanging over time*. The same coin cannot continue to be used if it is wearing out, for example. Moreover, the data gathering becomes invalid if the coin flipper begins to develop a skill in flipping coins. Absolutely nothing can change over time.

    Stated another way, the coin and the coin flipper must be independent of time. If the events are *not* unchanging (that is, if they are not independent of time), then every possible combination of coin flips will *not* have the freedom to potentially occur. This too is an accepted premise in statistics.

These two slight details of statistics are by themselves not contro-versial. They are accepted as being factually true. As we demonstrated earlier in the computer experiments chapter, when we model these two conditions on a PC, and run the experiment over and over, we ob-tain the same result, the same shape of distribution curve, every single time.

The resulting curve—a histogram of the averages of the binomials (heads/tails, yes/no coin flips)—is bell-shaped. Always.

When these two conditions are met, the existence of a bell-shaped curve centered at 50 percent is as dependable as the existence of objects falling to the ground. All statistics programs that model the two condi-tions of independence and unchanging over time produce completely replicated random distributions of numbers—that is, distributions that always have the same bell shape.

If everyone who uses statistics knows that this is true, then isn't the question answered? Isn't the case closed? The answer, plain and sim-ple, is no. Somehow we keep forgetting to ask the following founda-tional questions:

Do physics and nature meet these two key conditions?

Are real events in the physical world completely independent, and are they unchanging over time?

Think about this: What if *nothing* is actually independent in na-ture? What if, instead, all things in nature are interdependent to vari-ous degrees? In other words, what if nature does not meet the first condition to achieve a random distribution? What if the first key is not present?

Moreover, what if *"no things"* in nature are actually unchanging? What if, instead, all things in nature are changing in various degrees over time?

In other words, what if nature does not meet the second condition to achieve a random distribution as well? What if the second key is not present?

The reasoning is straightforward, the logic inexorable. If nature does not meet the two key conditions to achieve randomness, then ran-domness cannot occur in nature. Consequently, if nature does not meet the two key conditions to achieve randomness, then we cannot use the random sampling explanation for the existence and evolution of order in the universe.

It is worth repeating that if nature does not meet both key conditions to achieve randomness, then logic dictates that science must search for an explanation other than random sampling to understand the origin and evolution of order in the universe.

UNDERSTANDING INVISIBLE FIELDS—GRAVITY, INTERDEPENDENCE, AND THE EXISTENCE OF THE G.O.D.-PROCESS FOR SIR ISAAC NEWTON

Let's consider invisible gravitational fields. It is an accepted fact that gravity exists. Even very young children play games based on their acceptance of gravity. And gravity per se definitively establishes that everything is interconnected to various degrees. Yes, definitively.

Sir Isaac Newton was the first scientist to specify, both experimentally and mathematically, that gravity is the "invisible glue" holding the universe together. His famous mathematical formulas describe how everything—that is, every *thing* that has mass—has an associated gravitational pull that extends out into space in all directions. And this is true of all objects. An apple falling from a tree is not simply pulled down toward the earth. The apple pulls the earth toward it, and it pulls up toward the tree, and up higher still toward the moon, the sun, and beyond.

The apple is no different in exerting these pulls than the earth, moon, or sun; it is simply smaller and less compact. A physicist would say the apple has less mass and therefore has a less intense gravitational field. The earth pulls on the moon, and the moon pulls on the earth. Both the earth and the moon pull on the sun, and the sun pulls on both. In turn, the earth, moon, and sun pull on the apple, and the apple pulls back. The force the apple exerts is miniscule in comparison.

Newton was not only a mathematician and a scientist; he was also a sophisticated mystic and a firm believer in an intelligent G.O.D. process. In his time, to demean him, it was said that Newton held a simple clockwork or machine/mechanism vision of the universe. That myth has filtered down and follows Newton's reputation even in our world.

But the facts are quite different. Newton's integrative vision was deeply spiritual as well as experimental, but the scientific community was unwilling to acknowledge the fundamental parallel between Newton's idea of gravity and his vision of "God." It is important to understand that during Newton's era, science was being separated from religion. A successful separation was essential if science was to evolve. However, in the process, the scientists who championed the separation distorted some fundamental truths. I believe it is time to correct these distortions.

Newton reasoned that all objects or things, by definition, have mass, and all masses have gravitational fields that function as invisible attracting forces that extend out into space, in all directions, connecting everything that exists with everything else, to various degrees. These invisible attracting forces are completely replicable and trustworthy—they are always present.

Try to imagine these invisible fields extending out into space in all directions. Ponder how they interconnect millions or billions of stars in a swirling spiral galaxy, and further how they interconnect millions or billions of galaxies in a swirling, expanding universe. The thought is positively daunting. Physicists have a terrible time trying to model what is called the "three-body problem"—can you conceive what the interconnecting network of fields must be to model a "trillion-body problem"?

The gravitation field is completely unconditional—it pulls on any and all objects that have mass, and it makes no judgments. It does not matter whether the given object is large or small, black or white, good or bad. What matters is that the object has mass. If it has a mass, it has a gravitational field, and it will pull on everything else in the universe in varying degrees. To summarize, the invisible gravitation field functions as a mutually attractive, unconditional, nonprejudicial, and completely dependable force. Gravity is the universal force that literally holds the universe together and enables it to exist as a single, unified system—a "uni-verse" or "one-story."

What does it sound like to you when we envision the existence of a fundamental force that is universal, powerful, attractive, unconditional, and nonprejudicial? What does it imply to you when science discovers the existence of an infinitely interconnecting invisible field that unifies and holds the entire universe together? Newton viewed the integrating force of the universal gravitational field as being a physical expression of the existence of a "God" who was not only unifying but was universally loving, too. Is there scientific justification for Newton's vision that the existence of a universal and unifying gravitation field points to the existence of what I am now calling a G.O.D. process?

As we've seen earlier, *if everything is interconnected to various degrees by the universal gravitation field, so that nothing is completely independent of every other thing, then the first key condition for random sampling—that it be completely independent of other influences—is not met. Once this is understood, it follows that the distribution of planets, suns, galaxies, and networks of galaxies cannot be bell-shaped and will therefore* not *appear random. If objects did not have gravitational fields, there would be no galaxies and no superclusters of galaxies. There would simply be a random distribution of stars and planets throughout the vacuum of space.*

If you digest this one concept and incorporate into your very being, it will enable you to see everything with new eyes.

The conclusion that follows is that it is possible to observe complex and replicable orders in the cosmos because the universe is highly interconnected, by the gravitational field as well as by other invisible fields. This conclusion is abundantly confirmed; it fits exactly what we see. Astrophysics documents that galaxies are typically spiral shaped, for example. They are not randomly shaped. We now know that even the distribution of galaxies in the universe is not random, but reflects a complex arrangement or network of macro structures that look surprising similar to the network of cells in the human brain.

According to astrophysicist Sten Odenwald, Ph.D., author of *Patterns in the Void: Why Nothing Is Important,* even the seeming "vacuum" in space is teaming with fields of energy that are patterned or structured—that is, nonrandom. And the take-home message is that the gravitation field does more than hold the universe together. It contributes to the creation of macro orders in the universe. The existence of gravitational fields removes from nature the first key condition that is essential for permitting any randomness to exist in the universe. Importantly, the gravitation field is not alone; the universe is teaming with interconnecting fields operating from the quantum level, through the electromagnetic level, to the gravitational level.

THE QUANTUM FIELD AND SUPERINTERDEPENDENCE

Einstein's disagreement with the early quantum physicists in the 1920s and 1930s stemmed in part from his opinion that something was fundamentally wrong with how quantum physicists were interpreting Heisenberg's uncertainty principle. Whereas Heisenberg interpreted the observations of "unpredictability" as meaning "uncaused" and "unplanned," Einstein interpreted the observations as reflecting an as yet hidden set of complex organizing processes.

Einstein accepted the existence of some degree of freedom and choice in the universe. He recognized that some events in nature might be too complex for humans to predict precisely, but he was intuitively (as well as emotionally and intellectually) opposed to the conclusion that events at the quantum level had to be totally random (as produced by random sampling) and completely independent of one another.

Einstein's intuition and reasoning turn out to be accurate. Quantum physicists, as a rule, make an assumption that is fundamentally inconsistent with the very foundation of quantum physics itself. On the one hand, most physicists recognize a level of connectedness—and interdependence—at the quantum level that defies both the speed of light as well as common experience. Replicated physics experiments have documented that under certain conditions, two photons—once together and now traveling in opposite directions at the speed of light—continue to behave as if they were one. They behave as if there were no distance between them—a surprising conclusion that defies common sense (to put it mildly!).

The mathematics predicts that when two such photons begin together and then simultaneously travel even to opposite ends of the universe, what happens to one photon can be precisely and instantly observed in the other. This is called Bell's Theorem (no relation to the bell-shaped curve). The phenomenon is called entanglement.

This law isn't just theoretical; it has been confirmed in the lab. If an experimenter changes the spin of one of a pair of previously connected photons, the other photon will react seemingly instantly in the same way. This observation is termed "nonlocality," which means that distance or space does not matter when photons are completely connected—assuming the speed of light is fixed, as Einstein hypothesized. Despite the incomprehensible distances that separate them, they appear to react synchronously. Talk about interconnectedness!

Yet, on the other hand, many quantum physicists assume that randomness (implying random sampling) must be the explanation for why they are unable to predict simultaneously both the speed and momentum of quantum events—a phenomenon predicted by Heisenberg's uncertainly principle. Quantum physicists assume that these events are random because they have not integrated the logic of probability statistics with the observations of quantum physics. Yet logically there is a deep disconnect between the reasoning used by physicists to understand quantum physics and the logic used to apply statistics to data. When we repair this disconnect, the implications touch virtually every discipline of science and life.

I well understand this profound conceptual disconnect. As a scientist, I too was educated to reason this way. However, the experiments and logic reviewed in this book strongly suggest that it behooves science to revisit this conceptual disconnect if our true understanding of the universe is to evolve. Interdependence is the rule and not the exception in nature, especially at the quantum level.

The message is clear. The macro world of gravitational fields and the micro world of quantum fields both lead to the same conclusion: everything in nature is interconnected and interdependent to various degrees. And it follows that since there is no independence, then there is no support of a randomness theory, at least in its purest sense, as posited by random sampling. This conclusion is unavoidable.

SYSTEMS, FEEDBACK LOOPS, NETWORKS, AND COMPLEXITY

Because everything is interconnected to various degrees, everything is changing over time to various degrees.

A system is a set of components that mutually affect one another. In the process of components joining together, interacting, and sharing information and energy, the system changes over time. The inherent property of systems to change is termed "dynamical." Dynamical systems develop novel properties that are termed "emergents"; the novel properties "emerge" because the components interact and affect one another. Anything that exists in a dynamical system is constantly changing over time through interdependence and feedback. This fact has profound implications for the existence and evolution of complex orders in the universe.

Emergent interactions happen at every level of nature. For example, protons, neutrons, and electrons come together to make atoms. Both hydrogen and oxygen have the same components—protons, neutrons, and electrons—but when they become atoms, their emergent properties are unique. Moreover, the properties of both hydrogen and oxygen are different from the individual properties of protons, neutrons, and electrons examined separately. Hydrogen and oxygen come together and make a molecule called water. Water is remarkably different in its characteristics from the individual properties of hydrogen and oxygen. From subatomic particles to molecules, to biochemicals, to cells, to organisms, to ecologies, to solar systems, to galaxies, to superclusters of galaxies and beyond, novel emergent properties appear—at every level, and everywhere.

There is nothing random about the emergent property process. Though it is highly complex, it is highly replicable. When hydrogen and oxygen join together they make water, not gasoline or maple syrup. When human eggs and sperm join together they make people, not plants or protozoa. In the process of becoming a system, the components must interact—the components therefore become interdependent and constrain one another. Moreover, they constantly interact with systems outside themselves as well. Water interacts with temperature, for example, and it also interacts with the container that holds it. If the temperature is decreased below zero degrees centigrade, and the water is sitting in a heart-shaped dish, the water will solidify into the shape of a heart.

Feedback plays an essential universal role in these processes. The neural networks of the brain learn through neural feedback. People learn social behavior through social feedback. If we take away the neural feedback from a neural network, the neurons will not learn. Similarly, if we take away social feedback from people, people will not learn social behavior. Feedback is what enables learning to take place in systems, regardless of their level. And all systems, at all levels, have feedback.

All systems store information and energy to various degrees. The logic is straightforward—if a system has feedback, it will change dynamically over time to various degrees. Feedback loops constrain systems—the output of the system feeds back to the input of the system and controls the system accordingly. Feedback provides guidance in systems. What is termed positive feedback establishes the capacity for increasing complexity in systems, and in the process, it removes the potential for randomness to operate.

TURNING RANDOMNESS ON ITS HEAD

Let's review the essentials. When a scientist accepts the reality of fields in physics and quantum physics, and integrates this information with systems science and complexity theory, it is possible to discover that there is no longer a valid reason to assume that (1) events function independently in nature (the first key condition for random sampling to occur) and (2) events do not change over time (the second key condition). In a word, nature does not meet the two key conditions for randomness—as predicted by random sampling—to occur.

Yet even a youngster can flip coins and get an average of 50 percent heads and tails. And we can create computer programs that can generate bell-shaped curves 100 percent of the time. Something is missing if they can appear random. What happens when we create the two conditions that approximate randomness? I have already demonstrated that when the two conditions that approximate randomness are created, what happens on the average is actually completely predictable. We do not observe the origin and evolution of complex orders—we observe the same average order every time.

So, how can Einstein's statement be true? What did he mean? I suggest that we reframe Einstein's famous phrase slightly to say that the G.O.D. process can make it possible, within limits, to play dice in the universe—where "dice" allows some degree of creativity and freedom to express itself in nature.

Remember, the origin and evolution of the complex orders we witness in the real universe cannot be created by random sampling pre se. Throwing dice can neither create a universe nor sustain it any more than throwing sand in the air can create a sand painting or sustain it.

As we have said, in a field-interconnected universe—from quantum fields to gravitational fields—the conditions to create random sampling are for the most part not met. Therefore, it is not valid to use random sampling as a theory to explain the origin and evolution of complex orders in the universe.

There are, however, certain select circumstances of relative independence in the universe, and they can serve a very intelligent purpose. If a higher intelligence—some sort of Guiding-Organizing-Designing process—is required to produce the amazing complexity of evolving orders witnessed in the universe, we must ask why this superintelligence would make it possible to flip coins or generate numbers seemingly unpredictably? What higher purpose would the existence of seeming flexibility provide for the universe, and for us? Is this an expression of intelligent evolution?

The answer is simple. Creativity and individuality require relative independence. In order to have relative flexibility—and for systems to have relative degrees of freedom and choice—there must be relative independence. Simply stated, relative independence does not create novel orders per se by chance; what relative independence does is give novel orders the chance to occur.

That's important enough to repeat: *What relative independence does is give novel orders the chance to occur.*

What chance means in this case is *opportunity*. The fact that *we* cannot decode a given pattern of events or predict how the events will unfold over time does not necessarily mean that there is no pattern to the events or that the events are unpredictable. Einstein experienced aspects of the universe as being "incomprehensible." His opinion was that certain superorganized patterns were merely "incomprehensibly complex." And this book comes to the same conclusion. It takes a superintelligent process to create the kind of apparent intelligent evolution we observe in the universe.

Hidden in what is currently perceived to be randomness are supercomplex patterns that are waiting to be appreciated and ultimately discovered by man's evolving experimenting minds. Humans have discovered whole numbers, prime numbers, irrational numbers, calculus, Boolean algebra, matrix algebra, nonlinear differential equations, and infinity mathematics. It is worth remembering that everything is evolving, including mathematics.

Could the capacity for relative randomness be part of G.O.D.'s plan—a gift of gifts that enables us, and everything else, to evolve with relative creativity, semiautonomy, and sacred responsibility? Is G.O.D. a universal mathematician and sacred scientist? The journey continues.

*I see a certain order in the universe and Math is one
way of making it visible.*

MAY SARTON

INTERLUDE

The "Divine Proportion"

WHY G.O.D. IS A MATHEMATICIAN

Probably the most remarkable and revered number—which happens to be termed an *irrational* number by mathematicians—is also the most ubiquitous number in nature. The number is termed *phi*, and like its brother *pi*, the order of its digits is mysteriously unending and seemingly nonrepeating.

Here is how the astrophysicist Mario Livio, author of *The Golden Ratio*, introduced his readers to "the world's most astonishing number."

> *What do the delightful petal arrangements in a red rose, Salvador Dali's famous painting "Sacrament of the Last Supper," the magnificent spiral shells of mollusk, and the breeding of rabbits all have in common? Hard to believe, but these very disparate examples do have in common a certain number or geometrical proportion known since antiquity, a number that in the nineteenth century was given the honorifics "Golden Number," Golden Ratio," and "Golden Section." A book published in Italy at the beginning of the sixteenth century went so far as to call this ratio the "Divine Proportion."*

For most people the explanation of the phi ratio is easy to follow but impossible to understand. Consider the easy part first. Imagine you have a line that extends from A to C. Approximately two-thirds of the way along the line, we pick a point and call it B. We now have two segments—A to B (the longer segment), and B to C (the shorter segment)—that together equal the total line length, A to C.

Now, it's possible to adjust the position of B in such a way that the ratio of the longer segment (AB) to the total length of the line (AC) equals the ratio of the shorter segment (BC) to the longer segment (AB). (You may want to read that sentence again!) In other words, we can adjust the position of B so that AB divided by AC equals BC divided by AB (for those of you who like formulas, it is AB/AC=BC/AB). This is the simplest way to describe the golden ratio mathematically.

The phi ratio, like the well-known pi ratio (the circumference of a circle divided by its diameter), cannot be calculated precisely. No matter how far mathematicians carry out the division, the numbers keep appearing in a seemingly nonrepeating and unpredictable fashion. They appear to be random, save for the fact that every time the ratio is calculated, the identical sequence of numbers appears. What's truly astonishing about the phi ratio is that it fits diverse shapes and dynamics spanning art and architecture, botany and biology, physics and mathematics, and even economics and astrophysics. The golden ratio is found to occur in the dynamics of the stock market as well as the arrangement of billions of stars in a spiraling galaxy. How can a ratio that produces an irrational number containing a seemingly infinite series of never-repeating digits be the basis of a universal shaping process that extends from the Small to the All?

Here is how Dr. Livio asks the question: "Is the universe by its very nature mathematical?" Another way to frame this is in the famous words of the British physicist Sir James Jeans (1847–1946): "Is God a mathematician?"

A superintelligent G.O.D. process is guaranteed to be supermathematically gifted as well. As people evolve in their mathematical skills and understandings—and learn to see beyond apparent randomness—it is possible to take a quantum leap in comprehending the universal nature of the Experimenting G.O.D. mind. If that's possible, we will all reach Einstein's dream of coming to know the mind of God.

> *There is certainly no scientific reason why God cannot re-*
> *tain the same relevance in our modern world that He held*
> *before we began probing His creation with telescopes, cy-*
> *clotrons, and space vehicles.*
>
> WERNHER VON BRAUN, ROCKET SCIENTIST

11

Why Science Shaves with Ockham's Razor

SIMPLE PREDICTIONS THAT RADICALLY
TRANSFORM OUR MINDS

Does the fact that chance, by itself, cannot explain the origin and evolution of order in the universe require that people accept some kind of superintelligent designing process to explain everything that exists in the evolving universe? The key word here is "everything," and the commonsense answer is "Of course not." Even if you accept the fact that sand paintings never paint themselves, or that watches never assemble themselves, you surely recognize that many phenomena in nature seem to organize themselves quite well.

Commonplace experience reminds us that clouds appear to emerge spontaneously, raindrops appear to liquefy and fall naturally, and tornados appear to swirl and create destruction, all seemingly by themselves. Oak seeds appear to grow into oak trees, and fertilized human eggs appear to grow into newborn babies, again all seemingly by themselves. Phenomena that apparently invent and evolve themselves are scientifically termed "self-organizing systems." Nature is replete with systems that show apparent self-organizing properties.

To show you how to see the seemingly obvious alternative explanation to an intelligent, invisible Guiding-Organizing-Designing process with clarity and understanding—and I emphasize the words "seemingly obvious"—has been the goal of the book to this point.

Let's now pull together the three core principles previously discussed, to recognize the important implications that emerge from the totality of both the evidence and the reasoning presented so far.

PRINCIPLE 1: INTERPRETATION DOES NOT EQUAL OBSERVATION

In Chapter 9 I noted that the sun appears to revolve around the earth because we implicitly interpret our earthbound observations of the sun as if the sun were revolving around us.

It is, indeed, possible to make mistakes when interpreting one's observations. But sometimes a mistake is colossal, as when it was taken for granted that the sun revolved around the earth, whereas the opposite is actually the case. A fundamental lesson from the history of science is that things are not necessarily as they appear to the untrained, uninformed eye.

Yes, raindrops seem to form by themselves. And oak trees seem to grow by themselves. This is how it seems. But the earth also seems to be flat. And the sun appears to revolve around the earth. The underlying message must be respected—a commonplace interpretation of an observation, even a completely replicated observation, cannot be assumed to be necessarily accurate or correct.

This principle applies to all relationships in all systems, including the relationship between the mind and the brain. Since this book is ultimately about intelligence, including intelligent trial-and-error learning and evolution, it is instructive to discover how the observation/interpretation principle provides clarity and understanding to the fundamental question "Does the brain create the mind, or does the brain serve as a receiver for the mind?" Did the hardware create the software, or is it the other way around?

If an interpretation is consistent with the evidence—a subject we'll come to shortly—then the question arises, "Was the brain (the hardware) created by a superintelligent mind (the ultimate software) to serve as a tool of the human mind?" Neuroscientists interpret three kinds of evidence to support the interpretation that the brain creates the mind. The evidence *seems* compelling. But is the interpretation correct?

The three kinds of evidence are:

> *Evidence from Recordings.* Scientists can record brain waves (EEGs), using sensitive electronic devices. For example, researchers have discovered that alpha waves over the occipital cortex in the back of the head decrease when people see visual objects with their eyes and, as well, when people imagine visual objects with their minds.
>
> *Evidence from Stimulation.* Scientists can stimulate various areas of the brain, using electrodes placed inside the head or magnetic coils placed outside the head. For example, experiments have shown that stimulating the occipital cortex typically causes people to experience visual sensations and images.
>
> *Evidence from Ablation.* Observation of people and animals whose brains have been damaged through injury or disease, or who have had various areas of the brain surgically removed, have revealed that when areas of the occipital cortex are destroyed, people, dogs, cats, rats, and mice lose aspects of vision.

The commonsense—and generally accepted—interpretation of this set of findings is that visual experience is created by the brain.

However, the critical question is whether this is the necessary and only interpretation of the findings. The correct response is, certainly not. I first learned this lesson in logic as a young child when I took television sets apart. I then learned the lesson formally when I was an electrical engineering student at Cornell.

How do television sets receive their images? Televisions function as receivers for information carried by certain external electromagnetic fields. As we all understand, the sets do not create the visual information—they detect it, amplify it, and display it.

What most neuroscientists seem to forget is that the three kinds of experiments they conduct are the same three kinds of experiments that television repairmen conduct!

Evidence from Recordings. TV repairmen can record the television signals by using sensitive electronic devices (sensitive probes attached to digital oscilloscopes). For example, they can place the probe on particular components in the circuit and correlate the signals or readings shown on the digital oscilloscope with the visual images seen on the TV screen.

Evidence from Stimulation. TV repairmen can stimulate various components of the television, connecting electrodes from a device at appropriate points inside the television. For example, they can stimulate particular circuits with specific patterns of information and see replicable patterns on the TV screen.

Evidence from Ablation. TV repairmen can remove various components from the television (or areas can be damaged or wear out). For example, they can remove key components, and the visual images on the screen will disappear.

Do these three sets of evidence necessarily imply that the origin of the TV signals is inside the television? Of course not. They tell us that both brains and televisions play a role in visual experience—*they do not tell us what type of role.* The three kinds of evidence, by themselves, do not tell us whether brains or televisions are (1) self-creating the information or (2) serving as complex antennas-receivers of the information (which comes from outside the systems).

In other words, the three kinds of evidence do not allow us to determine whether the signals—the fields—are created inside the system (the interpretation applied to brains) or are coming from outside the system (the interpretation applied to televisions). Clearly, additional kinds of experiments—for example, signal-blocking experiments, using electrical and magnetic shields placed around the television set or the entire person—are required to determine whether the system in question (a brain or a television set) justifies a creator interpretation of the observations or an antenna-receiver interpretation of them.

In the case of brains, the fact that they seem to be involved with conscious experience does not necessarily mean that they are creating consciousness. Consciousness may require invisible external fields. Similarly, in the case of self-organization, the fact that objects appear to self-organize does not mean that they are actually organizing themselves. Organization may require invisible external fields as well.

Finally, in the case of evolution, just because growth and increases in complexity appear to occur naturally does not mean that evolution is non-intelligent. *Evolution—and what scientists term natural evolution—may require invisible external fields modulated by intelligence, that is, intelligent evolution.*

PRINCIPLE 2: INVISIBLE ORGANIZING FIELDS ARE REAL, AND THEY ARE THE RULE IN PHYSICS

Physics tells us, in no uncertain terms, that invisible fields—including gravitational, electromagnetic, and quantum fields—are the rule, not the exception, in nature and the universe. We live in a veritable sea of invisible fields. This infinite sea of fields permeates the vacuum of space with incomprehensively complex networks of structures and organizations. As astronomer Sten Odenwald writes in *Patterns in the Void: Why Nothing Is Important,* "We are forced to confront the fact that something hidden in the Void is controlling not just the subtle properties of matter but the destiny of the universe."

As Einstein said it, "The field is the only reality." All material things—including living systems—are organized by fields, as well as generating them. In light of this compelling evidence, logic requires that we entertain the hypothesis that invisible fields play a fundamental role in all physical phenomena observed in nature and the universe.

PRINCIPLE 3: WHY USE TWO EXPLANATIONS WHEN ONE WILL DO?

The fourteenth-century teacher and logician William of Ockham is remembered for a principle he set down that we know as Ockham's Razor; it was made popular in Carl Sagan's novel and movie *Contact*. Sagan provides a valid statement of the principle when he has his central character, Dr. Ellie Arroway (played in the film by Jodie Foster), make the remark, "All things being equal, the simplest explanation that accounts for the largest amount of the evidence is usually the correct one."

As Ockham originally stated his principle, *Frustra fit per plura, quod fieri potest per pauciora,* which means, "It is vain to do with more what can be done with less," and *Entia non sunt multiplicanda praeter necessitatem,* which means, "Entities should not be multiplied unnecessarily." Scientists as a group tend to be conservative; they seek "parsimonious" explanations in an effort to be conceptually frugal. I tend to be this way too. Here's how one could apply Ockham's Razor to the two sets of evidence that need explaining:

Evidence Set 1: Certain phenomena have never been observed to organize spontaneously (for example, sand paintings, watches); the evidence is clear that they require some sort of an intelligent Guiding-Organizing-Designing mechanism for them to come into existence, assemble, and evolve.

Evidence Set 2: Other phenomena have been observed to organize seemingly by themselves (for example, tornadoes, oak trees); they appear to self-organize when the right conditions are present.

One approach—the single interpretation approach—is to posit that a common invisible intelligent organizing field process is involved in both sets of evidence. (An organizing field process can operate externally or internally—since fields exist both inside and outside individual material systems.)

A second approach—the double interpretation approach—is to posit that an external organizing process (for example, a visible or invisible Guiding-Organizing-Designing field) is required for Evidence Set 1, and that some sort of internal organizing process (termed "self-organization," which in biological systems is governed by an internal plan/code called DNA) is required for Evidence Set 2. In other words, it's possible to employ two different explanations to account for observations.

However, to apply Ockham's Razor honestly, it would be best to search for the simplest explanation that accounts for the largest amount of the data. By definition, this would be the single interpretation approach. The G.O.D.-field interpretation can explain both sets of evidence, whereas the self-organization interpretation can explain only one set.

Which is your preference—one explanation, or two? If you prefer economy and adhere to intellectual integrity, then it follows that there are both philosophical and practical reasons for seriously entertaining a single invisible G.O.D. process in nature. Still, what ultimately matters is not which is the most economical (the G.O.D.-process interpretation), but rather which one is true. Ockham's Razor doesn't always lead to the the right answer; sometimes the less economical, more complex interpretation actually reflects the truth. Only with additional research can theories that make new predictions eventually be confirmed or refuted.

But it turns out that Ockham's Razor is sharper than might be expected. It is possible to put the invisible G.O.D. universal field process to some novel and revealing experimental tests and, in the process, discover some potentially hidden mechanisms underlying the "destiny of the universe."

AN ANCIENT VISION OF THE SUN AS BEING AN ORGANIZING EVOLVING SYSTEM

Would you be willing to consider doing a simple thought experiment with me? If so, then imagine that a huge meteor hit the earth, creating a cloud of dust that has blocked the sun's energy for many years. Imagine that the temperature drops globally to the point where all water freezes into a solid mass. Would we continue to observe rainstorms? Would we continue to see oak seeds grow? When water is frozen, water-related self-organizing processes would be observed to stop self-organizing. And now what do you conclude from this observation?

An obvious conclusion is that the many so-called self-organizing systems are not *completely* self-organizing and actually require the presence of other factors, including external fields of energy provided by the sun. A relatively narrow range of temperature is required for water to exist in a liquid state, which is necessary for rain to fall and for oak seeds to sprout and become trees.

But some more subtle and amusing interpretations of the sun's potential role in apparent self-organization should be mentioned.

The sun is a complex dynamical system that emits a wide spectrum of energies and information—from micro and radio waves, through infrared waves, light waves in the visible spectrum, and ultraviolet waves, to X rays and gamma rays. And all of these bands of photons carry dynamically fluctuating patterns of energy and information that can be thought of as being nonrandom quantum codes (as Heinz Pagels called them). These complex electromagnetic and quantum codes can play a communication role in modulating physical systems on the earth and all of its inhabitants, including us. We are all highly interconnected with the sun's bands of energy and information.

The sun's energies can be imaginatively described as a Beethoven symphony played by a world-class orchestra. The complete spectrum of dynamically changing information of the sun, from the lower subvisible frequencies, through the intermediary visible frequencies, to the higher supravisible frequencies, can be thought of as ranging from contrabasses (containing subaural frequencies), through violins, to piccolos (containing supra-aural frequencies), playing together full-time in a 24/7 orchestra.

The similarity between the sun and a symphonic orchestra in terms of generating organizing field information and energy can be shown experimentally. For example, it has been demonstrated that if a thin layer of sand is placed in a square pan and the pan is placed on top of a speaker, sounds of different frequencies will vibrate the pan, creating replicable patterns of hills and valleys in the sand—which can be described as simple sand sculptures.

A number of questions may come to mind: Many people wonder if it's possible that essential codes for life are carried on the dynamical spectrums coming from the sun. Others ask if it's possible that what we see as self-organization on the earth is made possible, at least in part, by the dynamic formative fields that are continuously emitted by the sun. And does the sun create morphological fields that help guide, organize, and design what we experience as dynamic physical life? Moreover, is the sun itself a gigantic antenna-receiver-transmitter of invisible intelligent G.O.D.-field signals?

If you favor Ockham's Razor and the search for integrative-unifying explanations and interpretations, then you can see how the G.O.D.-field process interpretation not only integrates diverse information but *makes new predictions that can be confirmed, or disconfirmed, in future research.*

I repeat: Ockham's Razor in this case not only integrates information in a highly parsimonious way, but makes new predictions that can be confirmed, or disconfirmed, in future research. This is one of the important reasons why Professor Wigner would likely label my G.O.D.-field process theory as "most amusing."

DO WE TRUST OCKHAM'S RAZOR? CAN WE "S.O.S." AND "SAVE OUR SOULS"?

We will never discover whether a G.O.D.-field process is actually interacting and communicating with us unless we are prepared to listen for it. Whether it is acting through nature, through the sun, or ultimately through the complex dynamic network of organizing fields that comprise everything in the universe—including the "void" or "vacuum" of space that is teaming with organizing energy and information—we'll never find out unless we go looking. And unless we learn how to interpret these dynamical information codes, we will never know what the codes actually mean.

Consider the following example: "Dit dit dit—daa daa daa—dit dit dit" may sound like a collection of ordered sounds with no meaning until we learn that, to the old telegraphers and ham radio operators using Morse code, the pattern "dit dit dit" stood for the letter *S* and "daa daa daa" stood for the letter *O*. A memory aid for the emergency signal "S.O.S." was "Save Our Ship." According to this book, perhaps that should be revived and revised to stand for "Save Our Souls."

If we are going to heal and transform ourselves and the world, then perhaps the time has come for us to open our minds to the possibility that the answers to our most pressing questions are all around us. It's possible that the answers to these questions are being broadcast, like gravitational fields, in all directions—ready to be received and adopted by intelligent and motivated individuals. This help in the form of universal fields of dynamic information available to each of us may be waiting to be received and interpreted.

To help you remember this hypothesis, you can think of this S.O.S. signal as coming from the S.E.L.F.—the Supreme-Eternal-Living-Field—which is a universal expression of the G.O.D. process. From this perspective, the S.E.L.F. is involved in the creation of all organized systems; hence, what we see as self-organization is a special case of S.E.L.F. organization. This is the G.O.D.S.E.L.F.

Ockham's Razor and its favoring of the intelligent-evolution explanation helps open our minds to this seemingly incomprehensible yet comforting possibility.

Part Five

IMPLICATIONS OF G.O.D.

FOR EVERYTHING

Three questions worth considering . . .

If the evidence for some sort of a G.O.D. process is virtually beyond any doubt, what implications would this have for how we live our personal lives and how we structure our educational, science, business, political, and religious institutions?

Is the human capacity to foster both personal and world healing dependent upon our ability to awaken to this higher reality?

Can science take people to a vision of G.O.D. in terms of Universal Consciousness that expresses the true genius within?

Imagination is more important than knowledge.

ALBERT EINSTEIN

12

Who Are We and Why Are We Here?

YOUR MIND IS BIGGER THAN THE ENTIRE UNIVERSE

Before we can appreciate the extraordinary capacity of the mind to guide, organize, and design, we must come to understand the profound potential of the mind to envision and imagine. The capacity of the human mind to envision its apparent infinite potential is as vast as the universe itself. Moreover, the mind is ultimately beyond anything we have yet witnessed in the physical universe. The mind's inherent potential not only goes beyond anything it has seen to date in the physical world; its potential goes beyond anything it can currently imagine. It can even be demonstrated through simple experiments that your mind's potential is bigger—in fact, much bigger—than the universe itself.

The human mind, along with its brain, has the ability to discover an infinite creative and organizing force in the evolving universe— that is, a G.O.D. process—because we are ultimately made of the same stuff and have the same infinite potential as "it." Unfortunately neither you nor I can appreciate who we really are, and discover why we are here, until we come to recognize this fundamental truth.

Again, this wisdom is captured in that aphorism "The All is in the Small."

I chose with a clear purpose the words "infinite potential" to describe the human mind. The history of philosophy and science reveals the human mind as being able to envision the infinite and furthermore as being able to develop a formal mathematics of multiple infinities even though humans cannot precisely picture what a single infinity actually is, much less a plurality of them. Just because we precisely picture something doesn't mean we can envision it. Envisioning includes the capacity for metaphor and symbols—a power that science is just beginning to understand.

It turns out that it's quite simple for science to demonstrate the capacity of the human mind to envision the concept of infinite, and then to go beyond it. As I take you through this purposely playful yet profoundly meaningful mental demonstration, I want you to imagine yourself performing this thought experiment with me. There is no substitute for having the direct experience yourself. In the process of experiencing this fundamental truth, you will realize that you can share this demonstration with your children, and help them celebrate the infinite power of their growing minds.

HOW OUR MINDS CAN EXTEND FROM THE "INFINITELY SMALL" TO "BEYOND THE INFINITE ALL"

Please take a quarter or other coin and place it in your hand.

We are about to take an unprecedented mental journey in eleven discrete steps—the size of the first nine steps ranging in scale from the infinitely Small to beyond the infinite All. We will jump from subatomic physics and chemistry, through biology and psychology, to ecology and cosmology, and beyond. The tenth and eleventh steps contain two surprising take-home messages.

STEP 1. A MASSLESS AND SIZELESS PARTICLE IN YOUR HAND

Imagine that the coin you are holding is a single photon of light. What does it look like to you? Is it bright? Is it pulsating? Is it changing colors? Is it heavy? Physics tells us that a photon is presumed to be a massless or "virtual" particle (it doesn't have any weight at all) and it is infinitely small (it doesn't have any size at all, either). Whatever you can imagine is okay for now. Just realize that you can imagine what a photon *might* be, and that physics experiments can be conducted to explore what part of your guesses might be correct. But not now. Remember, infinitely small photons are presumed to reflect the building blocks of everything in the known physical universe. As the Bible says and contemporary physics supports, the universe began with light.

STEP 2. A SINGLE ATOM IN YOUR HAND

Now, imagine that the coin you are holding represents something much, much bigger. The quarter has become a single atom of hydrogen; for the sake of time, we are skipping the intermediate steps of subatomic particles such as quarks or electrons and protons.

What does the atom look like to you? Is it solid? Is it pulsating? Is it emitting photons you can experience as colors? Can you see the nucleus in the center of the hydrogen atom? Try to see the electron cloud surrounding the nucleus. Again, whatever you can imagine is okay. The important thing is to realize that you can imagine what hydrogen might be, and that chemical research can be conducted to explore and confirm or contradict your images.

STEP 3. A COMPLEX BIOCHEMICAL IN YOUR HAND

Now, let's imagine that the coin you're holding represents something that is again much, much bigger. Once more skipping many levels, imagine that you are now holding human DNA in your hand. What does it look like to you? Does it have two strands? Is it shaped like a helix? Can you see the billions of bases that compose this essential biochemical molecule of life? Whatever you can imagine is okay.

What's important is that you appreciate that you can imagine what DNA might be, and we can conduct biochemical experiments to explore it.

STEP 4. A SINGLE CELL IN YOUR HAND

The coin you are holding is about to get much bigger again, jumping many levels of size. This time imagine that you are holding a cell, a human neuron, a single cell of the nervous system. What does it look like to you? Can you see the dendrites? Can you see the synapses? Can you imagine the neuron firing? Whatever you can imagine, the key is that you can try to picture the neuron.

STEP 5. A LARGE ORGAN IN YOUR HAND

The coin is now about to make a jump in size that will be particularly dramatic. Imagine that you are now holding approximately 100 billion cells in your hand, in the form of a human brain. What does it look like to you? Can you see its shape? Can you see the ripples in its surface that look like valleys? Can you see the brain pulsating? Can you imagine electrical brain waves flowing across the surface?

STEP 6. A WHOLE ORGANISM IN YOUR HAND

The coin is about to increase in size again. Imagine that you are now holding an entire person in your hand. What does it look like to you? Is she a child, or an adult? Can you see her hair color? Is she smiling at you? What is she trying to tell you? The key is that you imagine this person.

STEP 7. A ROTATING PLANET IN YOUR HAND

Once again, a huge jump in size is about to take place in your hand. Imagine that the coin you are holding represents the entire planet earth. What does it look like to you? Can you see which areas are cloudy and which are clear? Can you see the deserts and the oceans? Is the earth spinning on its axis? How heavy does it feel? We can conduct ecology experiments to discover the nature of this planet.

STEP 8. A SWIRLING GALAXY IN YOUR HAND

The jump in scale for the coin you are holding in your hand will now become even more extraordinary. Imagine that you are holding an entire galaxy in your hand! Can you see the billions of stars that are like tiny cells in its structure? Can you see its spiral shape? Is the galaxy spinning in your hand? How heavy does it feel? Imagine the galaxy, and we can conduct astronomy experiments to discover the nature of this almost comprehensively large system of stars that you are holding.

STEP 9. THE ENTIRE UNIVERSE IN YOUR HAND

All these steps of imagination—from photons and biochemicals, through cells and people, to planets and galaxies, are but training for what you are about to do. You are about to go beyond what anyone has ever seen—beyond what the most sensitive telescope has witnessed. What I want you to do is to imagine that the coin in your hand has become the entire universe! You are literally holding everything that exists in the entire universe in the palm of your hand. What does it look like to you? Can you see the billions of galaxies cupped in your palm? Is the universe itself pulsating? Can you see it expanding? Is it spiral shaped, or is it like a ball? Is it evolving, becoming more complex as you observe it? Whatever you can imagine is okay. However, it's not possible to test scientifically what you are experiencing, because what you are holding in your hand—imagined through the power of your mind—extends beyond the most sensitive of astrophysical devices yet conceived of by the human mind.

STEP 10. YOUR MIND HOLDING THE UNIVERSE IN YOUR HAND

As you have seen, the coin can be imagined as a photon, infinitely small, or as the universe, infinitely large. There is nothing to stop you from imagining that the coin even contains an infinite number of universes, each having an infinite size. The fact is, whatever you can envision and imagine, you then have the potential to go beyond. Theoretical mathematicians do this all the time.

And next—this amazing step; along with a powerful question. Who is it who is *aware* that his or her hand is holding the universe? Who are you who can imagine holding an evolving universe or universe of universes in your hand? Clearly, the "I" that is "you" can look beyond virtually everything—and I mean everything. Your mind can witness all of this, and more.

The potential of your mind is so vast that it is beyond comprehension—but now, as a result of this innocent visualization exercise, you have some sense of its vastness. Over these ten steps, you have used your mind to hold every level of everything that has ever existed in the universe in the palm of your hand. The question is, where does all this mind power come from? Did the mind just happen by chance? Or is your mind part of something else? Can you imaging holding your mind in your hand?

STEP 11. THE UNIVERSAL MIND OF GOD IN YOUR HAND

However powerful our individual minds are, we can raise the question "Is there a Universal Mind that is beyond ours?" Are our individual minds—our "small" minds—part of a much greater mind, the Universal Mind of the All? And does the Universal Mind of the All exist within our small minds? As we have asked throughout this book, is the All within the Small?

Can you imagine, as you hold the universe in the palm of your hand, and "you" have this experience, that a mind even greater than yours is witnessing you do this? This may be a new idea for you, but if you try, you can imagine this possibility, too. In fact, you have the inherent capability with your mind to imagine holding the Universal Mind in your hand.

So, even as you imagine beyond the All of the universe, you can imagine something even greater that is enabling you to do this. That is how gigantic the power of your mind is. You can actually imagine having the power of "God"—even holding the entire universe in your hand—while, simultaneously, you can imagine that you are a living expression of an infinite G.O.D. process.

AN ESSENTIAL CAVEAT

Those of us formally trained in clinical psychology, psychiatry, or social work have had experience treating mental madness. We know well that just because it's possible to imagine something does not mean that the something is real. And although science fiction frequently precedes science discovery, some science fiction is precisely that—fiction. However, these criticisms are obvious, and they do not undercut the essential message of the exercise you've just been through.

Which is greater in scope: the currently known physical universe, or the human mind that contemplates it? "Who can hold what in one's hand?" Yes, the universe is outrageously larger than our material body. This is an obvious physical fact. However, *the scope of our mind is extraordinarily larger than the known physical universe.* This is a less obvious psychological fact but a fact worth remembering, and celebrating. It is the key to awakening to the power of our minds and their role in the cosmos.

APPRECIATING THE EVOLVING MIND IN AN EVOLVING UNIVERSE

The truth is, no one knows what the actual potential of the mind is—because whatever greatness we have already achieved as a species suggests that there is no apparent limitation to the mind's ultimate potential.

Facing reality in the twenty-first century causes one to ponder the scope of contemporary evolution as we witness it. When I use my word processing program on my laptop computer, in the righthand corner of my screen sits a little wizard who turns his head, blinks his eyes, raises his hands, and every now and again nods his head in approval. Sometimes he even raises a tiny wand in his hand. He reminds me of the amazing imagination of the human mind and the remarkable evolution of computer software over the past thirty years.

I can switch to my Internet browser with the click of a mouse, and immediately receive vast amounts of information about virtually everything—from superstrings and photons, through cells and people, to astrophysics, metaphysics, and science fiction. All this knowledge and information is a manifestation of the ever-evolving human mind.

Over thirty years ago at Harvard, I purchased a PDP11/GT40 computer system that came with 16K (16,000 bytes) of memory and cost $40,000. The computer took up a whole room. I purchased an additional 8K of memory that was the size of a large suitcase, at the cost of $5,000. This large machine did not have calendar, phone and address book, word processor, or spreadsheet software. I had to program the computer using Fortran, BASIC, or FOCAL to draw a graph or put text on the screen.

My PDP11/GT40 computer, though it cost a great deal of money, didn't have a little blue and gold Merlin with a big nose, mustache, and beard. The little Merlin reminds me of the stupendous power of the human mind to both understand and shape the world, and in the process, understand and shape itself.

Today I carry a pocket digital assistant with 64 megabytes—64 million bytes of memory. It remembers my schedule, phone numbers, addresses, favorite quotes, ideas, writing plans, questions to ponder, expenses. It is a genuine mind assistant and was designed by human minds. I paid $500 for it in 2002.

What technology will exist thirty years from now? Will the jump be as great as over the last thirty years? What do you think?

The desk I had been using when I wrote this chapter was near a window that faced the entrance of the Metropolitan Museum of Art in New York City. Hanging from the upper ledge of the museum was a two-story green banner, probably twenty feet wide and thirty-five feet long, imprinted in huge white letters with the word GODDESS; it was meant to attract New Yorkers to the current art show. The potential of the human mind is certainly showcased by the work of great artists who, through recorded history, have envisioned goddesses and gods.

I began to wonder, is the need to visualize goddesses and gods simply a reflection of human ignorance and superstition combined with a heavy dose of arrogance and wishful thinking? Or does our need to imagine some sort of a G.O.D. process actually reflect our implicit understanding that our true potential genius as an emerging species expresses the infinite genius of a conscious intelligent evolving universe?

Are we humans like little Merlins—tiny goddesses and gods—gifted not only with personal and family responsibilities, but planetary and cosmic responsibilities as well?

As you now have learned, chance will never create a Michelangelo painting, let alone a simple sand painting or a little painting of Merlin. People create paintings. What is painted is a reflection not merely of what already exists but of what can exist in the future.

THE EVOLUTION OF DESIGN AND OUR EVOLVING UNDERSTANDING OF DESIGNERS

Although this book offers evidence that leads inexorably to the conclusion that some sort of cosmic Guiding-Organizing-Designing process exists in the universe, I want to step back for a moment to consider different ways of envisioning what a designing process is. Clarity of thought is necessary in order to appreciate how this book's conclusion is dependent on the scientific evidence. If we are to ever conceive and understand intelligent evolution, we must become intelligent about evolution, and therefore become intelligent about the process of design.

There are at least five different ways of envisioning how design operates on the earth and may operate in the universe as a whole.

The first and simplest way is to posit that each designer functions independently of every other. Independent design creativity is viewed here as completely individual, reflecting the unique propensity and personality of the designer. Most thoughtful people recognize that this simple perspective, by itself, is incomplete. Humans do not exist or learn in a vacuum—we learn from one another, as well as from nature.

This leads to the second and third ways of envisioning the designing process. Here we assume that designing is governed partly by shared genetics (gene-mediated design) and partly by shared experience (environment-mediated design). In other words, our capacity to design—as well as the particulars of what we design—reflects our common genetic heritage plus our common environmental heritage. Whatever we design—

automobiles, buildings, computers, drugs, games, jewelry, music, poetry, software, trains, weapons—reflects a combination of our DNA code plus our respective experiences and education.

These three ways of envisioning the designing process—independent, gene-mediated, and environment-mediated—though obviously true to various degrees, reflect only part of the designing process. A fourth way of envisioning design can be termed "interdependence." The fact is, typically we do not design as individuals; we design as groups, teams, or systems. From software to automobiles to weapons, teams of tens, hundreds, or thousands of individual minds are brought together to design things that work. "The whole is greater than the sum of its parts" applies to the designing process and the evolution of designs over time.

Those who are open to the existence of a larger spiritual reality—as described by the evidence reported in my book with Bill Simon, *The Afterlife Experiments*—will be able to envision the possibility of a designing process that includes team members "from the other side." While serious investigative work in this controversial area has only just begun, scientific integrity and responsibility still require that we include the possibility that a designer or design team could be influenced from this other side source as a fourth way of understanding design. However, the question arises, are these four ways of envisioning design sufficient to account for all aspects of design—not just by us, but as it is expressed all around us? The evidence from Parts Two through Four says "No."

Based on the evidence I have presented, it is beyond reasonable doubt to conclude that all things existing—from subatomic particles, through humans, to galaxies and beyond—have not occurred by chance. Therefore there is some sort of a fifth evolving creative design process, which I term G.O.D. It is also beyond reasonable doubt to conclude that humankind's capacity to design things, as well as the particulars of what we design, reflect a purposeful manifestation of this universal designing process.

HOW DOES THE G.O.D. PROCESS CONTRIBUTE TO OUR INTELLIGENT DESIGNS?

What is the reasoning that leads us to conclude that our personal efforts at designing are guided, at least in part, by a higher if not universal designing process? There are many reasons why I include a fifth way of envisioning human designing, as reflecting the expression of a universal, macro, or meta-designing process in the universe. Here are a few of them.

First, as you now appreciate, complex orders in the universe cannot come about simply by chance—they require some sort of an intelligent designing process. Hence, the premise that some sort of universal Guiding-Organizing-Designing process exists everywhere in the universe is a required logical deduction.

Second, if everything that exists in the universe is a reflection of this universal creative designing process, then by definition the "everything" includes you and me and all of us. It logically follows that if universal design is the rule and not the exception, then humans are an integral expression of this designing process. We apparently carry the latest evolution of the designing process, at least in terms of what is known on the planet at this time. Simply stated, if any evolved organism on this planet has the capacity to serve as an antenna-receiver for the universal G.O.D. process, it is the human organism.

Third, it's well known that most of the energy and information that science has documented as existing in the universe cannot be directly experienced through our five senses. What we see with our eyes—as exquisitely complex and beautiful as it is experienced through vision—reflects only a tiny number of all photons in the universe. The visible frequency spectrum of light is but a tiny slice of the entire electromagnetic/photonic frequency spectrum of energy discovered and accepted by current science.

Invisible fields affect the functioning of every cell in our body to various degrees. By definition, the All of energy includes the complete set of known energies in the universe, as well as those yet to be discovered. We should resist the temptation to be arrogant and conclude that current science has uncovered everything there is to discover about potential energies and evolving fields in the universe. The existence of dynamic invisible fields provides one plausible means by which a universal designing process could potentially orchestrate infinitely complex unfolding plans and designs.

The evolution of cell phones reminds me of the vast potential for invisible communication and guidance in the universe. I recently attended a University of Pennsylvania graduation ceremony and witnessed something I could never have imagined forty years ago when I was an undergraduate at Cornell. There were approximately three thousand graduating seniors sitting in chairs in Franklin Field, and there were probably more than fifteen thousand family and friends in the stands.

While the faculty was marching in, everywhere I looked on the field I saw undergraduates speaking into cell phones, and everywhere I looked in the stands I saw family and friends speaking into cell phones. There were probably more than a thousand simultaneous cell phone conversations taking place between the field and the stands at that very moment. The radio waves were of course invisible to the naked eye. There were no wires connecting the students to their parents and loved ones. The calls were interconnected via the invisible radio frequencies. Most remarkably, each of them could hold a separate conversation—they did not experience one massive jumble of frequencies and words!

It's valuable to remember that humans did not create radio waves; the universe did. However, humans have learned how to use radio waves for long distance communication and control. Is the existence of radio waves, and their remarkable properties, merely an accident of nature that humans have harnessed for special purposes? Or do radio waves exist for one or more purposes provided by a Guiding-Organizing-Designing process? Could radio waves be a manifestation of intelligent evolution?

ARE PHOTONS PURPOSELY DESIGNED TO BE "WEIRD"?

Sir James Clerk Maxwell, the distinguished nineteenth-century scientist—who believed deeply in God—was the first researcher to provide an explanation and set of mathematical formulas that integrated the entire electromagnetic spectrum. His formulas led the way for scientists in the twentieth century to discover how all electromagnetic waves, including radio waves, could function as photons. And, as mentioned before, most photons vibrate at frequencies that are invisible to our physical senses.

Yet there is a curious property of photons—including the visible ones—that for years struck me as particularly odd until I asked myself the question, If photons were designed this way, what was the purpose of the design? The word "odd" is a modest term for this truly amazing fact of photons. A more honest term is "beyond extraordinary." Many people who learn this fact will make a point of never forgetting it. Moreover, if one were looking for a phenomenon that deserved the label "miracle," it would be this property of photons.

If in a vacuum we take one laser beam, and then cross it with a second laser beam, an astounding phenomenon occurs. When the second laser beam is turned on and it reaches the first laser beam, if the two beams are completely in sync, the streams of photons will summate when they cross. If the two beams are completely out of sync, the streams of photons will cancel each other out when they cross. But after they cross, the beams continue *as if nothing had happened at all.* The two beams cross, they combine or cancel, and then after they cross, they return to their prior state *as if nothing had happened.*

Simply stated, visible light is "invisible" to light. Though the statement is simple, the explanation remains a mystery. While the totality of the light—termed "interference"—is momentarily altered when the light beams cross, the individual beams are completely transparent to each other: *they do not lose their individuality once they have crossed.*

This phenomenal property of photon action raises a fundamental question. Answering this question requires that we go beyond the current comprehension of science. How can one beam of light have its energy add or subtract from another beam of light, and then continue as if nothing has happened? How is this physically possible?

The interpretation provided by contemporary physics leaves one's head spinning in disbelief—yet may be correct. We are told to accept the idea that photons are massless "no things" that are "spaceless" and function metaphorically as "ghostlike somethings." Photons are described as "infinitely small," and we are told they vibrate with utter accuracy, and because they are massless, they are invisible to one another. And this is what physics asks us to accept without proof.

Why does light behave in this curious way? Why would the universe create invisible massless and spaceless quanta of energy with this bizarre property?

I conclude that the reason for the bizarre quality of light is communication—communication of both existing and potential orders and designs!

A simple way to picture this is as follows. Imagine that you have two children—Skott and Kim (the names of my godchildren)—and they are standing a few feet apart from each other. The light that reflects off Skott reaches both of your eyes. The light that reflects off Kim also reaches both of your eyes. If you doubt this, try closing each of your eyes in turn as you look at two children in a row. You can see both children with either eye. The light from Skott and Kim must therefore cross in complex ways as they reach each of your eyes. If photons were like billiard balls—objects with mass—then as the objects bumped into one another, they would change their direction and behavior. By the time the billiard photons reached your eyes, you would see a mess—a mixture of the two children.

However, photons do not function like billiard balls—they function like virtual particles. Because they are completely invisible to one another, they can convey the individual shapes and behaviors that capture the individual structures and functions of Skott and Kim. So we can see two people at the same time as distinct beings because of the extraordinary properties of light. Is it reasonable to speculate that this extraordinary property of photons happened by chance alone?

When I look at the sky at night—especially at the big sky over Tucson—I can see thousands of stars with my naked eye. Through powerful telescopes at the Kitt Peak National Observatory outside of Tucson, I could see millions of galaxies, each containing billions of stars. And with the Hubble telescope located in space, I would be able to see billions of galaxies. All this light crosses in the vacuum of space. The light from billions of billions of stars crosses in the vacuum of space, yet each and every star maintains its individuality. It can communicate its uniqueness. Try to imagine this fact!

Photons not only have a kind of immortality—which is how we can see light that has traveled for millions if not billions of years, but they also provide information with such precision and accuracy that incomprehensibly huge amounts of information can be contained in incomprehensibly small spaces.

If mere humans can design future space telescopes to look into the farthest reaches of the universe, isn't it possible that the universe—so to speak, a cosmic astrophysical G.O.D.—could have designed the nature of light itself in such a way that would make it possible for us to look into the farthest reaches of the universe? Remember, for the light from the edge of the universe to reach our eyes, and for us to see the images clearly, the light would have to function in such a way that it preserved the information carried over billions of years of time. Is it possible that "God's codes," as in Pagels's "cosmic codes," are expressed as information in invisible light beams? For example, could there be an infinitely complex code of information carried by cosmic superrays that penetrate all objects to various degrees?

Perhaps the universe acts like one huge cosmic cell phone. Perhaps the Guiding-Organizing-Designer messages are being broadcast throughout the entire universe. If so, perhaps everything that exists in the universe picks up this signal to various degrees.

Is the world awakening to this profound cosmic communication process? Are humans on earth meant to discover this process just as we discovered radio waves? Are we, as the philosopher Spinoza wondered, germinating intelligent cells in the evolving body of God?

The word "science" comes from the Latin *scire,* which means "to know." I believe that if such questions are not asked, answers may never be found.

Remember, you—and your children—have the potential, with your individual minds, to conceive of holding individual photons in one hand and simultaneously holding the whole universe in your other hand. You are meant to wonder and to know.

It has been said, "As above, so below." Systems science teaches that the reverse is also true. "As below, so above." The Small is in the All, and the All is in the Small. The implications—for science, education, business, politics, law, religion, and all of human society—are staggering.

Not everything that can be counted counts, and not everything that counts can be counted.

ALBERT EINSTEIN

13

Evidence-Based Faith

A NEW WAY OF MARRYING SCIENCE AND SPIRIT

Belief is profoundly important in everything we do. Belief is what guides our attention. Belief is what leads people to seek in particular directions, and therefore to discover. Belief also plays a powerful role in motivation. It's one thing to know how to fly an airplane; it's another to believe that you can do it. For whatever reason, knowledge is not enough. So it is my belief that science might provide definitive answers to fundamental questions such as "Does the soul continue after physical death?" and "Is the universe the result of a creative and intelligent designing process?" Belief in possibilities is what motivated me to conduct the research presented in *The Afterlife Experiments* and to proceed with research and analysis for this book.

What we choose to believe, what we ultimately come to accept on faith, comes from five possible sources. So that science and faith can become two sides of the same coin, I will now describe how it is possible for you to take a scientific approach and extend your belief to faith. I propose that if people could develop the capacity to practice evidence-based faith, we could have the potential to globally resolve our conflicting beliefs and faiths, and in the process discover the means to live in peace, balance, and harmony.

FIVE WAYS OF FORMING BELIEFS

In the following section, I use the term "belief." Once you understand how the five-part structure is applied to the word "belief," you will be able to reapply the structure to the word "faith." The potential transformation from belief to faith is important because to have faith one must have trust. When we witness things happening 100 percent of the time, such as apples falling from trees, implying the existence of invisible gravitational fields, or sand paintings mixing in a box, implying that something more than chance is required to create sand paintings, we can put our faith or trust in this evidence.

Evidence-based faith is far different from blind trust; it is open-minded discerning trust. And it is intelligent trust. Similarly, evidence-based belief is not blind belief; it is open-minded discerning belief. And it is intelligent belief.

How many of these five ways do you use in forming your beliefs?

1. EDUCATION-BASED BELIEFS

The first way we form our beliefs is to absorb what we are taught by our parents, community leaders, teachers, and clergy. Individually and as representatives of their institutions, these people instruct us about what we should believe. The beliefs we are taught may be true, partly true, or not true at all; they may accurately represent quality values and reality, or they may be quite removed from them. The extreme education-based believer says, "If my parents/president/teachers/clergyman taught me this is true, then it must be true."

2. EMOTION-BASED BELIEFS

However, many beliefs are formed by factors other than education and culture. We often develop personal preferences for specific beliefs: our emotions lead the way. Our beliefs may be wish-based rather than education-based—or they may be a combination of the two. Sometimes our beliefs are shaped by unconscious needs and desires, including anger and fear. The emotion-based believer says, "What I believe is what feels right to me. If it feels good, it's real."

3. EXPERIENCE-BASED BELIEFS

Personal experiences and associated intuitions can powerfully shape our beliefs. When our experiences match what we have been taught and even what we want to believe, the resulting belief is more firmly embedded. When our experiences do not match what we have been taught, or what we want to believe, we can become seriously confused. But the committed experience-based believer says, "I believe what I see and experience, regardless of whether you see it or not."

4. REASON-BASED BELIEFS

Logic can also be a powerful factor that shapes our beliefs, especially if we prefer the rational over the intuitive. It is curious that logic often leads to conclusions that are inconsistent with education, wishes, or experiences. This is especially the case among thoughtful superskeptics. They fully understand how logic is leading them to form new conclusions that are inconsistent with what they were taught, what they want to believe, or what they have personally experienced, and they will become anxious and angry even as they are convinced. They prefer not to be convinced; it hurts too much.

The extreme reason-based believer says, "If there is no flaw in the logic, then it must be true." The extreme reason-based believer would reject any new logic that doesn't match his or her own.

5. EVIDENCE-BASED BELIEFS

This is the empirical, "show me" approach to believing. It is the approach of the genuine scientist, the open-minded explorer who looks not only to personal experiments for evidence, but to experiments conducted by others that are convincing.

In the deepest sense, all children are natural scientists. Through intelligent trial and error, they discover what is safe and what is not, what works and what doesn't, what gets them scolded and what gets them praised. They compare their experiences with others, and grow accordingly. Learning about reality is first and foremost an evidence-based process. The extreme evidence-based believer says, in essence, "I ultimately don't care what I have been taught, what I personally want or experience, or even what logic tells me. What matters most is what replicable experiments in my life and the laboratory reveal."

IGNORING LOGIC AND EVIDENCE

People whose beliefs are based primarily on education, emotion, and experience tend to dismiss or ignore logic and evidence. This is a pre-scription for bias, prejudice, torture, war in the name of God, and even worse. However, the great prophets, seers, and mystics combined personal experience with evidence; they looked for replication and verification before they drew their conclusions. They were "experimenters" in the truest meaning of the term.

Unfortunately, just because one uses the term "scientist" or even "evidence-based" does not mean that one is doing so with integrity. Too often scientists choose the evidence they prefer—what fits their education, emotion, personal experiences, or preferred reasoning—and discard the evidence they do not like. This, of course, is completely unacceptable. If a person is going to practice evidence-based medicine, create evidence-based theories, or even establish evidence-based religions, then all the evidence must be fairly represented whether one likes the evidence or not.

EVIDENCE-BASED FAITH

The term "evidence-based faith" may seem a contradiction in terms—an oxymoron. Most people keep one bucket for everything supported by evidence, and a completely separate one for everything based on faith.

But there are others like me who regularly practice evidence-based faith, even in the conduct of science. This applies to drawing conclusions that are not only beyond reasonable doubt but virtually beyond *any* doubt.

When I take a day away from the university campus to teach the concepts of energy and force to high school students or to medical researchers, I perform a simple demonstration. I take an eraser, a book, a set of keys—whatever objects are handy—and holding out my arm, I release the object from my hand.

Every time I've released an object in this way, it has fallen to the ground. Even though I know, in principle, that this will not happen in the space shuttle when it is orbiting the earth, and even though I can imagine, in principle, that someone might one day be able with their mind alone to slow the object's fall or even suspend the object in midair, what happens when I release objects on the earth is that they fall to the ground with a thump, a plink, or a splat.

Releasing objects is as reliable as taking a sand painting and shaking it in a pot. The object falls every time. The sand painting mixes always. Can we put our faith in something that happens 100 percent of the time? I would say "virtually yes." What I mean by this is that even though the object has fallen 100 percent of the time, and the sand has mixed 100 percent of the time, this does not absolutely guarantee that it will happen this way in the future. Something *might* change.

Hence, in order to continue to be virtually 100 percent sure that the event will happen in the future, it is necessary to continue repeating the experiment every now and again. The experiment must continue.

This is why I am attentive when I drop things. It never hurts to check. My essential belief is that it is adaptive to live an evidence-based life, and to put one's faith in evidence-based conclusions—which by definition always leave us with a little room for doubt.

However, if the evidence says that something has happened 100 percent of the time in the past, then we have reason to believe, not with *absolute* certainty but at least with *virtual* certainty, that it will likely happen this way in the future. This is the strongest form of evidence-based faith.

If something happens approximately 45 percent of the time—which is about the average of shooting baskets from the floor for great basketball players—we can put our faith in the fact that in the foreseeable future, it will continue to average around 45 percent of the time. We can expect the great basketball players to average 45 percent shooting—some games will be less, and some will be more. However, we will continue to gather evidence in order to keep our faith in a given player's performance. He could get depressed, or begin to lose his skill, or suffer from a debilitating infection. Or, over time, dazzling young players might come along who would manage higher averages.

Evidence-based faith is reality-based faith. It is open-minded faith. It is flexible faith. It is discerning faith. It is intelligent faith. Comprehensive evidence-based faith takes evidence from a variety of disciplines such as physics, mathematics, psychology, and parapsychology, and when the observations all lead to the same conclusion, accepts the resulting conclusion as evidence-based faith.

In this book I have taken evidence from physics (sand mixing 100 percent of the time), from mathematics (the normal curve appearing bell-shaped 100 percent of the time), from psychology (the mind's capability to extend its understanding from the infinitely small to beyond the universe itself), and from parapsychology (certain individuals being able to dream about future months' events in advance) as together providing evidence of orchestration of complex personal lives. I have considered the implications of this combination of evidence honestly and forthrightly.

This is not a book to promote taking on faith what our parents, professors, politicians, or priests tell us to take; or taking on faith what our emotions tell us; or taking on faith what our personal experience tells us; or even taking on faith straightforward logic.

Rather, I am suggesting that we should put our trust—and therefore our faith—in replicable evidence. Then if and when the evidence changes, we can change our minds accordingly.

MY PERSONAL EVIDENCE-BASED FAITH

I have come, slowly but surely, to develop an evidence-based faith concerning the existence of a Guiding-Organizing-Designing process in the universe as a whole. I have come to develop an evidence-based faith in intelligent evolution. I did not come to this faith from my parents (they were agnostics, at best) nor did I come to this conclusion from my education (in fact, many of my distinguished professors were atheists).

I did not come to this faith from my emotions (they were conflicted at best). Like most scientists, I wanted to be accepted by my peers, and I was frightened to entertain intelligent-evolution hypotheses that were viewed as mistaken if not outright taboo by many of my colleagues. But for the sake of humankind and life in general, I wished there was meaning and purpose to life and a way to explain it.

My faith is first and foremost evidence-based faith. I did not initially come to this faith from my personal experience. I never saw an apparition of a man with a flowing white beard, or had mystical experiences I could label with confidence as coming from the divine.

But once I began to be open to the possibility of a universal G.O.D. process while I was at Yale, and I began asking the universe for information, I had experiences that were strikingly consistent with the idea of a universal G.O.D. process that, with time and logic, led me to accept the conclusion about the existence of God.

The set of logical principles that brought me to that conclusion came especially from contemporary systems science—the logic that leads to the idea of the All in the Small.

However, it is the weight of the evidence, spread over multiple disciplines and contexts, that together brings me to a conclusion I can trust: that some sort of universal, creative, and intelligent designing process is part of the very essence of the evolving universe itself.

EVIDENCE-BASED FAITH—IMPLICATIONS FOR OUR LIVES

How might we live our lives differently if we were to accept the evidence-based conclusion of the G.O.D. process as true?

If evidence-based faith were adopted, how might we evolve not only our personal lives, but the institutions we create—from science and education, through business and law, to politics and religion?

First, let's consider our personal lives. Since we would now be open to a larger plan and purpose in our individual lives as well as the lives of people we interact with, we would begin to become ever more aware of the coincidences and synchronicities that exist in our lives. By pondering these matters and searching for their meaning, we would search within ourselves, as well. And this search would also include our relationships with others. We would continually ask questions like: "What might this coincidence mean? What is the lesson I am meant to learn from this situation? How can I take advantage of this learning, this gift, or this disappointment, to benefit myself, my friends and family, and the world? We would learn to listen to the G.O.D. process and perhaps we would learn to hear, as well.

Over the years I have become ever more aware of the obvious as well as subtle indications of coincidence and synchronicity in my personal life. Also, I have reconsidered major and novel events that have happened whose properties I later learned had profound meaning.

I am planning to write a book that will be devoted to describing how all people can discover connections to the G.O.D. process in their daily lives, using my personal journey as a playful example of how this can be successfully accomplished in an evidence-based faith. However, to offer one example here, I will briefly share an event that happened in my life and the spiritual meaning that occurred to me only thirty years after the event happened.

It was late May 1966, two days before I was to graduate from Cornell. I was driving on the East River Drive in New York City with Jeanne, my wife of one year. I was twenty-one, she was twenty. It began to rain heavily; she grew nervous and asked me if I would pull off at the nearest exit. Sensing her anxiety, I took the next exit. The exit was flooded. In front of us, a stalled Jaguar blocked the lane. I stopped. Our 1965 white Opel—a tiny car with no frame to speak of—promptly stalled too. I then did something strange. In my head I heard a voice that said, "Put on your seat belts, now." I had never worn a seat belt, and I had never asked Jeanne to wear hers either. I didn't even know how to put it on. However, I had never heard that voice in my head before, and I was not about to argue. I started struggling with my belt while asking Jeanne to please put on hers, as well. She was surprised by the request but did what I had asked. About a minute later, a

Mustang came barreling off the freeway at about fifty miles an hour and smashed into the rear of our car. We were literally sandwiched between the Jaguar and the Mustang. Our car was totally demolished.

Jeanne's seat was ripped from the floor. I was later told that if she had not been wearing her seat belt, she would have been thrown through the window and killed, or at least have been paralyzed from the waist down. I received a brain concussion and suffered temporary memory loss. Later I was told that in the ambulance, semiunconscious, I kept screaming, "I have retrograde amnesia." It was a correct diagnosis. The next morning I awoke and was told that my wife was in a room down the hall with a severe broken back.

Meanwhile, not only did I have no memory of the car accident, I had no memory of being married! I knew who Jeanne was—we had met each other as freshmen—but I had lost memory for about three years of my life. To this day, I have almost no memory from a couple of hours before the accident to twenty-four hours after it.

Within a week after the accident, much of my memory had returned, including a vivid recollection of hearing that protective voice in my head. The car accident not only dramatically changed Jeanne's life and mine, but affected the lives of our parents as well. It contributed to suffering that extended beyond physical pain. Jeanne's mother, Rose, committed suicide a few months later, in part over the mental anguish about her daughter's condition, and I made up my mind to go to Harvard, which would put Jeanne near her father and sister. I always wondered what my life would have been like if the car accident hadn't occurred. Jeanne's mother might have lived. I probably wouldn't have gone to Harvard. If it were not for Jeanne's severe back pain, which lasted for more than twenty years and prevented her from lifting weights, she and I would probably have had children.

But even more interesting is the insight that came to me just a few years ago. I had never considered the profound implications of an obvious yet inexplicable fact of this accident that I had overlooked. Our car was completely destroyed. The trunk was protruding through the backseat; the engine was protruding through the front floor. Jeanne's car seat was held in the vehicle by her seat belt. And I had received only a small bump on the back of my head, erasing my memory of the accident. I had no broken ribs, no significant bruises, no cuts from the glass. Nothing. Ordinarily an accident like this would have buried the steering wheel in my chest. Ordinarily an accident like this would have badly mangled or even buried me.

So how and why was I spared? Was the accident simply that, an accident? Or was I somehow protected? Was this part of my life's lesson—to come to ponder the existence of possible invisible guidance and protection in our lives? The fact is, when the accident happened, I never entertained such thoughts. I originally interpreted the voice in my head to be my own voice—Jeanne was anxious, I got anxious, and I simply made a wise choice. Case closed.

However, as I finally began to ponder the physical reality that something quite anomalous had happened—first I had heard a voice, then I did something I had never done before: put on the seat belt. Yes, I suffered temporary memory loss, but I had been completely spared from total devastation.

No one had taught me to be open to the possibility that there is more to life than meets the eye. No one had suggested to me that the amazing properties of light I had learned in physics might portend the existence of a truly amazing universe whose potential is greater than any of us have yet imagined. It took me over thirty years to develop evidence-based belief in a G.O.D. process and the implications of intelligent evolution. Now that I believe in the possibility, I am able to understand things I didn't even ponder before. This example illustrates more than just seeing meaning, purpose, and patterns in our lives, or learning how to go with the flow. It also suggests that we should not be afraid to ask for advice and guidance. In short, I suggest that we would all be wise to stay open to receiving helpful information from a higher intelligence—even if you don't know or aren't yet convinced of the actual source.

Another remarkable example of extraordinary coincidence and guidance supported by hard evidence in physical reality is included in Appendix C. The experiences reported there will truly challenge your ability to see novel information with open and discerning eyes.

When a thought pops into your head, the fact is you really don't know where it has come from. There are many possibilities that can be considered. The thought could have come from your conscious mind, your unconscious mind, the mind of someone else who is physically alive (telepathy), the mind of someone else who is deceased (mediumship), a higher intelligent entity (perhaps angels), or even the G.O.D. process.

Regardless of the source, if you don't ask for information, you may not receive it. If you are not open to receiving information, you will likely not hear it. However, if you are open to hearing from the G.O.D. itself, you have an increased chance of hearing. Remember, in this part of the book we are dealing with operating from the perspective of evidence-based faith. We are concluding that "the All is in the Small"—that an intelligent G.O.D. is in everything, including us.

In the personal information front page of my PDA, I sometimes have the phrase "Ask God First." I do this to remind me to develop a new habit—to regularly ask the universe for assistance, and see what pops into my head. Of course, what pops into my head could be just my imagination—it could be inaccurate if not crazy. I won't know until I ask the question and see what I receive. Neither will you.

For the record, I am not suggesting that we blindly accept whatever pops into our heads and act on this information indiscriminately. Discernment is a prerequisite to intelligent, wise, and reality-based decision making.

We must learn not to blindly reject information that comes to us intuitively. If G.O.D. is inside each and every one of us, we are foolish to ignore this clear and deep well of wisdom.

Seeing evidence of a higher purpose and plan, and seeking guidance about this creatively unfolding plan, are two concrete changes worth making in our personal lives. Furthermore, we can all become more open-minded about this invisible infinite potential, and in the process we can become humble servants of the potential goodness and healing that exists. Becoming more open-minded and humble is fundamental to growing and evolving. Systems science teaches that emergent properties—unpredicted, unexpected events—are the rule, and not the exception, in systems.

My fondest example is the one I gave earlier: how hydrogen and oxygen, two invisible gases at room temperature, can join forces and become the water molecule, taking on a set of unprecedented and completely unpredicted emergent properties that are unlike any other liquid in the universe. Just looking at the properties of hydrogen and oxygen as gases alone could not predict that they could create a liquid which when frozen, would create an infinite number of complex crystalline structures termed snowflakes.

At every level in nature, when new systems emerge, "the whole is greater than the sum of its parts"—the pattern of properties is unique. The potential for novelty in the universe is incalculable.

Every now and again a great mind is able to make a prediction about potential. Probably the most famous is Einstein's realization of $E=mc^2$, which led to the inception of the atomic age. Very few physicists would have guessed that so much power could be held within such a little space. These days physicists are predicting that there will be more potential energy in a single cubic inch of the vacuum of space than represented by all the matter in the universe. How's that for a spectacular prediction!

In the same way that objects fall and sand mixes, it is a logical and evidential fact that the universe has infinite intelligence and infinite potential, so I use the terms "infinite intelligence" and "infinite potential" with clear purpose and much care. With evidence-based belief in the existence of a creative designing process in the universe, a person becomes more open-minded, discerning, and humble. And in the process, a person becomes ever more thankful to have been given a mind capable of appreciating all that the universe has to offer. We become more appreciative and joyful as we are able to experience blessings on a daily basis.

I find it as difficult to understand a scientist who does not acknowledge the presence of a superior rationality behind the existence of the universe as it is to comprehend a theologian who would deny the advances of science.

WERNHER VON BRAUN

14

Implications of Intelligent Evolution for Society

BRINGING G.O.D. TO SCIENCE, EDUCATION, MEDICINE, BUSINESS, LAW, POLITICS, AND RELIGION

When I was a child, we started each day in school saying the pledge of allegiance to the flag. I remember the line "One nation, under God, with liberty and justice for all." Some of my more mature friends said, "one nation, with liberty and justice for all."

As I look back at my early education, I realize that I began each day pledging something I did not understand. What did it mean to say, "One nation, under God"? Was I to take this literally? Now I wonder if we could expand this great thought to say, "One species, under God." Or, "One world, under God." Or even "One universe, under God."

Raised to be an agnostic, I did not know whether or not there was a God but I did not mind saying the word in the pledge. Then when I was older, saying the words "under God" seemed unscientific to me, if not downright superstitious.

However, I later learned that the founding fathers of the United States were not only great leaders and visionaries but were deeply religious and spiritual as well. On the back left side of a dollar bill is a pyramid that is topped by a shining eye. Most people—including Americans—have no idea what the religious and spiritual significance of this symbol is. Some claim that the symbol comes from the Freemasons and the Rosicrucians, that Benjamin Franklin was both a master Mason as well as a Rosicrucian, and that Thomas Jefferson also was a Rosicrucian. This was certainly never mentioned when I went to school.

Another claim is that it is derived from a secret society called the Illuminati (a sect that some believe Galileo may have belonged to). Supposedly they called it their "shining delta." The eye purportedly signifies the Illuminati's ability to infiltrate and watch all things. The shining triangle represents enlightenment. The triangle is also the Greek letter delta, which is the mathematical symbol for change, transition, transformation.

Curiously, the words underneath the pyramid say *Novus Ordo Seclorum,* which can be translated to mean "New World Order" or "New Secular Order."

In fact, "secular" could also describe this book. I have not discussed any specific religious beliefs or practices—Hindu, Jewish, Christian, Islamic, Rosicrucian, Illuminati, or other. This is not a book about religion or religious orders—it is about scientific evidence for the existence of God and what this could mean to our future evolution. I mention the United States dollar bill and the phrase "In God We Trust" to remind the reader that the idea of God was central to the formation of this country—and it was connected to the concepts of "liberty and justice for all."

But that subject has been discussed often enough by the media, by scholars, and by politicians in Washington, D.C. Let me rather consider some of the implications of an evidence-based-faith approach to God for some of the major institutions that define human culture.

Given the scope of the G.O.D. idea, it should not be surprising to discover that the concept has implications for every human activity.

RESURRECTING INTELLIGENCE AND G.O.D. IN SCIENCE

If science has the potential to solve the mystery of mysteries as well as serve the mystery of mysteries—as stated in the Prologue—then it has a sacred responsibility to do so. If some of science's current theories need to be amended or eliminated in light of new evidence, then this should happen. Science has the obligation to amend or eliminate a theory if it turns out to be incorrect. Unlike classical religions, which are faith-based and can sustain their beliefs regardless of evidence to the contrary, science is supposed to be evidence-based and must change its beliefs in the light of new evidence.

The fact that certain scientists defend their favorite theories and theorists is understandable, and in certain contexts is even lovable. Scientists are protectors of ideas, and they have their heroes. However, every scientist who has ever lived has been human, and humans make mistakes. Not one of us is perfect. For science to have integrity, it must acknowledge not only that our understanding as humans may be limited, but that the universe itself may be evolving, and therefore our understandings may have to evolve as well.

It is my view that scientists need to develop and extend their capacity for open-mindedness with humility. But the training of these traits is not emphasized in contemporary scientific education; I will work for changing this unfortunate reality.

There are various levels of implications of evidence-based faith in God for science. The simplest level is to reconsider the role of random sampling and randomness not only in the universe, but in whatever discipline the scientist is studying. If everything is interconnected to various degrees, and true independence is present only in very unique and limited situations (for example, photons crossing in the vacuum of space), then we need to spend time gathering newer statistics that account for the massive interconnecting networks of energy and information. Though nature will be seen as ever more complex, containing intricate implicit purposes and plans, the resulting new discoveries about the organizing nature of nature itself will reward future scientists beyond imagination. It will no longer be acceptable to blindly apply random sampling techniques to interconnected phenomena; statistical techniques derived from chaos and complexity theory will need to be developed.

Beyond this, scientists face a challenge on an even more complex level: to build bridges between disciplines—to create integrative disciplines that look for parallels across domains. This book, for example, integrates examples of evidence from physics, mathematics, chemistry, biology, psychology, and parapsychology. The evolution of integrative science will discover interconnections that can allow for the revelation of creative intelligent design in all levels of science. If scientists and religious leaders can together agree to adopt an evidence-based approach to belief and faith, the capacity for each to inform the other will be manifest. Einstein put it this way: "Science without religion is lame, religion without science is blind."

For the record, I should state what is hopefully obvious by now. If you asked me the question "Are you trying to use science to prove the existence of God?" my response would be, "Absolutely not. What I am attempting to do is to use the scientific method in an open-minded manner to enable God—if he/she/it/they exist—to prove his existence himself."

I am not a God thesis advocate; I am a *Veritas* advocate—Harvard's motto, the Latin word for "Truth." And I am attempting to use science to provide the "Light and Truth" called for by Yale. If G.O.D. exists, then I will honor the evidence, and be its faithful advocate. I am neither an evolutionist, creationist, or intelligent designist. If I am anything, I am a "truthist"—following the evidence wherever it goes.

As we've seen, the evidence to date overwhelmingly rejects chance alone as a plausible explanation of order and evolution in the universe, and strongly supports the conclusion that some sort of an intelligent G.O.D. process exists in the universe. Intellectual honesty and integrity require that science finally honor this evidence and recognize the probable reality of intelligent evolution.

Even systems science—which leads to G.O.D. through its focus on interconnection, interdependence, complexity, and emergence—can evolve when G.O.D. becomes part of the equation. Systems scientists like to say that the relationship between two components (be they two particles, two people, two planets, or two galaxies) is a two-way street.

However, maybe the relationship is actually a three-way street—with G.O.D. as the center lane. Moreover, maybe secular science (which assumes that nothing exists until proven otherwise—technically called the "null" hypothesis in statistics) needs to be balanced by "sacred" science (which assumes that G.O.D. exists unless proven otherwise). This leads to the creation of a balanced, integrative science that combines the advantages of both perspectives for learning and evolving.

RETURNING INTELLIGENCE AND G.O.D. TO EDUCATION

There is no bigger or more important idea than an intelligent Guiding-Organizing-Designing process. If this thesis is true, then it should be a foundation of future education. Children hunger to be taught that the existence of some sort of G.O.D. process is as fundamental as the existence of gravity. Since education will be improved when fearless children face the teacher, G.O.D. is best incorporated, when appropriate, into education. This will add a solid base to the rapidly expanding, ever changing, mysterious world of knowledge.

"When appropriate" means to suggest that science teachers need not have their students pray to God for laboratory experiments to come out right, or that schools should encourage language teachers to have their students pray to God that the right meanings and grammars pop into their heads when they make a translation. I am also not suggesting that science or history be reinterpreted from any particular religious point of view. What I am suggesting is that students learn about the existence of God with the same thoroughness that they learn about their society or history. If universal intelligent, trial-and-error designing is essential for understanding the origin and nature of life, love, and liberty, then these ideas should be taught to students.

The new G.O.D. education I am proposing is not the limited faith-based God interpretation promoted by any single religious group—Jewish, Christian, Islamic, whatever. As also proposed in *Tomorrow's God,* it includes core aspects of each but does not include the whole of any. Though it is wise for the education community to be moderately religion-phobic, it need no longer be G.O.D.-phobic.

BRINGING G.O.D. TO MEDICINE AND HEALING

There was a time when priests and shamans were scientists as well as healers. They communed with God (or the Great Spirit), retained and transmitted their wisdom and traditions, made fundamental discoveries about nature and health, and served in the roles of doctors and counselors. They practiced rituals, meditated, and ingested special plants and herbs to enhance their communication with the larger spiritual reality so as to achieve the noble goal of serving their communities.

When we accept the evidence for the existence of a ubiquitous G.O.D. process in everything, including us, this provides an increased rationale for contemporary health care providers—be they hi-tech surgeons or low-tech energy healers—to pray for their patients as well as to become open to receiving guidance and assistance from above (sometimes referred to as intuition from within). Does the health care provider who believes in G.O.D. and prays for his or her patients have more successful treatment outcomes? The healing process becomes viewed as a three-way street between the patient, the healer, and G.O.D.

Perhaps we can look forward to a time when the G.O.D. process becomes an active and accepted component of the healing process. We would look to G.O.D. to assist us in discovering new uses for herbs and plants as well as help in creating combinations of compounds whose emergent properties are greater than the properties of the individual components (from simple synergisms to completely new properties). We would use conscious intention and prayer in the creation of new compounds to enhance their potential effectiveness. Though suggestions such as these may sound far-reaching (if not far-fetched), we must consider such possibilities with an open and discerning mind if we are to discover how far the G.O.D. process hypothesis may extend into our daily lives.

CONSIDERING ESSENTIAL FEEDBACK FROM G.O.D. IN BUSINESS DECISIONS

The business community is not particularly known for its lofty ethics and values. Business—particularly big business—is viewed by some as an institution that is motivated by greed, self-interest, and profit focus, with little concern for the future of our species or the planet. The United States government implicitly encourages its business women and men to push the limits of regulation so as to earn as much money as possible.

The argument can be made that the more money companies earn, the greater will be the number of people who have jobs, the greater our collective affluence, and the greater the influence and power our country will have in the world. Moreover, the thesis is that if problems arise for people and the planet, the public will raise its voice, and both business and government will respond reasonably. In principle this philosophy is fine, and our nation has grown financially strong from this approach. However, compelling evidence exists all around us that the philosophy may not be working very well, especially concerning the health of our planet, and by extension, the health of individuals.

Consider the following visionary possibility. What do you think might happen if business leaders were to reexamine their principles and consider receiving guidance and direction from the universe? If we are an evolving product of an infinite intelligence—and I emphasize "if"—the evidence indicates that we have been curiously given substantial freedom to believe what we want to believe regardless of whether or not our particular beliefs match reality. We even have the freedom to believe or not to believe in the existence of an infinite intelligence that potentially provided this gift of freedom in the first place. Freedom to believe, for better or worse, appears to be a purposeful part of intelligent evolution.

We have a choice not to believe in G.O.D. And we can run our businesses without thinking about G.O.D. in the context of what we do, which is of course the way most businesses function. However, we also have the choice to believe in G.O.D. and express it. We can, if we so choose, take G.O.D. into account in the design and execution of how we conduct our business lives. Should we "ask G.O.D. first" when we make business decisions? Should we place this information on the boardroom table, and see how it fits with what we want to do? Are CEOs ultimately under the invisible guidance of G.O.D.? What if corporate CEOs were to ask for more overt guidance from G.O.D.? Would our world become healthier, joyful, more peaceful and fulfilling?

INTELLIGENCE AND G.O.D. IN THE COURTROOM

We have designed our legal system, and our laws, based upon our collective experiences as a species. Save for the Ten Commandments that the Old Testament describes being given to Moses on Mount Sinai, humans have made their own social rules and designed their own legal systems. At least that is how we tend to think about these things. Of course, if we entertain the idea that everything that exists reflects some direction from an invisible G.O.D. who functions from behind the scenes, we can look to the structure and evolution of our laws and legal systems and potentially see evidence of invisible guidance in the evolving process.

The United States' legal system, the one I am familiar with, is currently in something of a crisis. The strategy of opposing advocacy—prosecutor versus defender—too often places more emphasis on winning than on discovering truth and rendering justice. The stresses in the system are symbolized by the process of the O. J. Simpson trial and the Supreme Court's handling of the Bush-Gore Florida voting crisis.

What would happen to the legal system if G.O.D. was brought into the courtroom? What would happen if ideas for improving how the legal system works, as well as specific decisions themselves, included some input from G.O.D.?

I'm not suggesting that we go to the local oracle and simply follow what she or he says. What I am offering as a possibility is that in the future, teams of "evidence-based oracles" will be able to obtain replicable information under controlled conditions that can be verified—and that the information these teams gather would be willingly taken into account as an integral part of a legal decision-making process.

If all this information, this guidance and wisdom, is available, waiting to be received under the right conditions, then it seems prudent for us to consider attempting to discover if this prediction is correct. Wouldn't it be wonderful if our evolution as a bio-psycho-spiritual species includes our awakening to this infinite wealth of virtually free information?

ONE EARTH, UNDER GOD, WITH LIBERTY AND JUSTICE FOR ALL

If evidence-based faith is justified, and evidence-based oracles can be developed in the future, the potential exists to evolve our world to the point where politics is world politics rather than national politics—where world peace with prosperity can prevail.

Throughout recorded history, tribes have attempted to control other tribes, and nations have attempted to control other nations. The warring nature of people has resulted in more injustice and tragedy than probably any other aspect of our behavior. The fact is, children can be raised to love like Mother Teresa or hate like Saddam Hussein. Humans are gifted with the potential to revere nature or rape her. We have the capacity to be tender to others or torture them. Like the surgical knife that is neither good nor evil—it can be used to save a life or to take it—humans, too, are neither good nor evil. We can teach each next generation to promote love and compassion, or to foster hatred and terror. Humans have the potential to heal or hurt.

The question is, what potential behaviors do we wish to nurture? What kinds of humans do we wish to become? What is to be our destiny and our species' legacy?

In a free society, politicians ultimately serve the people—the power is in the people. And in societies that are not free, time and time again the people have risen up against oppressors and insisted that their needs and dreams be met. However, with power comes responsibility and accountability. What is our responsibility, and for what are we each accountable?

Imagine that the answers we need to these vital questions already exist. Imagine that the information is available if we are willing to ask. And imagine that the key to opening this door to wisdom and greatness simply requires that we adopt evidence-based faith. Can politics evolve from disinformation and excessive ego to integrity and humble leadership? Can we think beyond our states and nations to become one people living on one world, with one G.O.D.? As the deeply spiritual film *Dragonfly* states so succinctly, "Belief is what gets us there." In the process of writing this book, I have come to believe that the best is possible. If it is real, it will be revealed, if we are willing and ready to receive it.

THE RESURRECTION OF HONEST AND COMPASSIONATE RELIGION?

Though I will probably regret writing the following words, I must share them for the sake of integrity. If we look honestly at the history of humankind, the evidence overwhelmingly shows that religion has played both the most grand and the most horrific roles in our evolution. In the name of God and brotherly love, we have killed and maimed our human, animal, and plant families. On the one hand, we have built the most magnificent cathedrals, and we have written the most haunting words and music. But on the other we have tortured or killed millions of people, or in milder circumstances treated those who did not agree with us with disrespect and injustice.

My personal opinion is that religion has been humankind's greatest social success and its most miserable failure. Religion has supported the very highest ideals and also practiced the very lowest. Religions have splintered into competing factions of all shapes, sizes, and colors. There is no unified Buddhism, Judaism, Christianity, or Islam. Each contains hundreds of factions (depending upon how you count them).

And the factions get along with one another about as well as the major religions get along with one another—which is not very well at all.

Religions, essentially faith-based and not evidence-based, if they wish to be concerned with truth, must ultimately include scientific evidence to help shape their beliefs and ideals. To repeat the words of Wernher von Braun that appear at the beginning of this chapter, "I find it as difficult to understand a scientist who does not acknowledge the presence of a superior rationality behind the existence of the universe as it is to comprehend a theologian who would deny the advances of science."

Religion needs to become evidence-based if it is to survive as a viable system for helping humans reach their potential, individually and collectively. As the Dalai Lama writes in his book *The Universe in a Single Atom: The Convergence of Science and Spirituality,* "If scientific analysis were conclusively to demonstrate certain claims in Buddhism to be false, then we must accept the findings of science and abandon those claims." He goes on to say that no one who wants to understand the world "can ignore the basic insights of theories as key as evolution, relativity and quantum mechanics."

I am not proposing that the world must have one religion any more than I am proposing that the world must have one culture. I have come to the conclusion that the G.O.D. process actively stimulates and celebrates diversity. But we can become one family, and live in harmony on one planet, without having to eat the same food, listen to the same music, wear the same clothing, or even speak the same language. Our capacity to be different—to be individuals—is part of our special gift and is consistent with an experimenting G.O.D. process that has contributed to the creation of nature that contains both ladybugs and scorpions, angelfish and tiger sharks, cardinals and vultures, tabby cats and cougars. In nature, diversity is the rule, not the exception. The same rule applies to humans.

However, behind virtually all religions are certain fundamental spiritual ideas and values that ultimately connect us to our common source—a G.O.D. Can religions come together, accept the mistakes of their past, correct the mistakes of the present, and evolve accordingly into the future without dragging their past differences along like Marley's chain?

All religions have their shadow side; the question is, can they acknowledge it and grow from it? (I do not mean to stereotype when I raise the following questions; not all religiously conservative persons hold these views.) Can fundamentalist Jews admit that some of their priests misbehaved badly during the time of Jesus and some of their present rabbis are engaged in economic behaviors that are inconsistent with universal love and compassion? Can orthodox Christians admit that some of their priests behaved inhumanely during the time of the Inquisition, and that too many of their priests have recently engaged in sexual behaviors that are inconsistent with universal love and compassion? Is it possible for conservative Islamists to admit that hatred and revenge drove many of their religious leaders to support the horror of 9/11, and that, even now, some of their leaders engage in terrorist behaviors that are inconsistent with universal love and compassion? As mentioned earlier in this book, what is required is courage and humility combined in balance—the eagerness to learn, to celebrate contributions and virtues, while at the same time admitting to mistakes. The fact is, none of us is perfect, nor are our institutions.

As for the future—can people create an evidence-based-oracles association that includes oracles from all the major religions? Can we conduct sacred science research in the future that honors different traditions united by a common agreement to let the data speak?

It is not enough simply to believe in a God that fosters love and compassion for all; we know that blind belief is insufficient.

What seems necessary is to bring science to God and God to science, and establish the living reality of a ubiquitous G.O.D. process. We must learn to accept the limitations of our past histories and raise them to a new level. All people must see what we can become, and decide that this is how we prefer to live and believe.

Billions of people are currently searching for purpose and meaning in their lives and the life of the universe, even as the evidence clearly indicates that the choice is ours to make. Science no longer is taking God away; science is discovering God in every place it looks and bolstering our beliefs. Science need no longer be seen as an enemy of God; science is becoming his/her/its/their ultimate servant. Contemporary science is becoming the universal tool by which we can know God and reveal the great potential—the spiritual genius—that exists within and around us.

Science is enabling people not only to see the power of our individual minds; it is enabling us to discover the Universal Mind that provides the spark that ignites us all.

A CALL TO EVOLVE EVIDENCE-BASED FAITH

Just as it is possible to conceive, in principle, of an infinite series of infinite numbers, we can conceive, in principle, of an infinite intelligence allowing infinite potential.

What we do with this stupendous power of mind—this penultimate gift—is up to us. Will we become servants of universal goodness, beauty, opportunity, and justice? Will we evolve into a bio-psycho-spiritual species of which we can be proud? Will we develop our ultimate gift, the capacity for infinite love (discussed in the Epilogue)? The evidence is all around us—the choice is ultimately ours. And from the perspective of evidence-based faith, this potential—this grand choice—is within our grasp.

How do we change our minds about mind and G.O.D.? Can science help us with this seemingly impossible task?

Changing one's paradigm is not easy. Millennia passed before humankind discovered that energy is the basis of matter. It may take a few more years before we prove that wisdom and knowledge are the basis of—and can actually create—energy, which in turn creates matter.

GERALD L. SCHROEDER, PH.D.

15

The Organizing Mind

NEW DISCOVERIES SHOW HOW THE MIND ORGANIZES MATTER

As I was writing this chapter, Lonnie Nelson, now a Ph.D. but then a brilliant graduate student in the Department of Psychology at the University of Arizona, was completing his fifth and final experiment for his master's thesis. When I looked at the results of this experiment, my reaction was, "Okay, I give up." He appeared to have demonstrated that a person's state of mind leads electrons in a special electronic device to become more organized, even at a distance. That conscious intention can increase the organization of electrons flowing in a resistor shielded from both electrical and magnetic influence is a concept worthy of courageous thought.

The device he had been using, a random event generator, or REG, detects electron "noise" in a resistor, which is assumed to be random. A computer counts and plots the distribution of this electron noise over time. The REG used in our laboratory was designed and built by the Princeton Engineering Anomalies Research laboratory (known as

PEAR) in the Department of Electrical Engineering at Princeton University. The pioneering work from PEAR, described in Professor Robert Jahn's and Brenda Dunne's book *Margins of Reality,* requires that we revise our understanding about how the universe works.

An elegant series of studies conducted over two decades has shown conclusively that the human mind can influence the seemingly random behavior of balls and electrons, not only from a few feet away but from a distance of thousands of miles away as well; in these tests, the human subjects demonstrated their influence when working with mechanical machines such as pinball-like devices, and with electronic machines such as REGs. They are also called RNGs, which stands for random number generators.

For most subjects, the effects are very small, but they are highly reliable and replicable. If the average number of hits and misses detecting an electron in the REG averages 50 percent, people can shift the average up and down by intention alone approximately 2 percent.

In Paul Pearsall's *Wishing Well,* current research on the organizing mind is reviewed, and the implications for human life and health are illustrated. When I read Paul's book, I remember feeling (and resisting) the Margaret Mead statement quoted earlier, "These are the kind of data I wouldn't believe, even if they were true." But these data are valid. The data are real, and truth is supported by evidence. Moreover, the data are predicted by contemporary physics and systems science.

It was Lonnie's fifth and final experiment that convinced me he had discovered something wonderful in our laboratory. One of the members of Lonnie's master's committee wanted to make sure that the effects he was observing were not due to subtle hand movements. To address this valid question, Lonnie and I designed the following experiment.

The subject was requested on some trials to move his hands up and down a few inches directly above the REG device. Though the device was electrically and magnetically shielded, and therefore should not respond to simple electrostatic body-motion field effects, we ran the hand movement trials nonetheless. On other trials, the subject moved his hands up and down four feet away from the device. It is known that electrostatic body-motion field effects decrease dramatically with distance; hence, if moving hands did have an electromagnetic effect on the REG, the effect would be larger when the hand movement trials were close to the device. The order of trials was counterbalanced as required by controlled experimental design. After each one-minute trial, the subject rated how absorbed he was (that is, how well he had managed to concentrate on the task), and also the degree to which he lost track of time (a measure of "trance" or "daydream" state).

The data spoke for themselves. Distance per se had no measurable influence on the average number of hits and misses of the electron's behavior detected and tracked precisely on the computer. In other words, the electromagnetic shielding worked as designed by the electrical engineers at Princeton. However, the trials where the subject reported being in a deeper trance or daydream state, regardless of distance, were associated with statistically significant increases in the organization of the electrons' behavior. Just as Lonnie had found in his previous four experiments, high states of trance or daydreaming were associated with increased organization of electrons in the REG.

It is important to understand that in this research, the subject was not attempting to influence the device per se with his mind. The person was simply moving his hands, or playing a computer game, or deliberately daydreaming, or doing different kinds of meditation. Regardless of the task, when the subject was more absorbed, the REG responded with increased order. In other words, in a system where electrons have relative freedom to change their behavior (that is, they have relative independence—recall that contemporary physics tells us that nothing is completely independent), the electrons spontaneously became more organized in the presence of an absorbed, organized mind. Remarkably, the organization of the subject's mind was paralleled to some degree by the organization of electrons in the REG. Mind and electrons, connected over distance, spontaneously resonating together. A new form of the synchronization described in Steven Strogatz's book *Sync*.

FROM ONE MIND TO MANY MINDS
ABSORBED IN SYNCHRONY

If you follow the reasoning, you are led to the prediction that groups of minds, joined by a common cause and absorbed together, might have an even greater effect on REGs. Research by Roger Nelson and colleagues from Princeton and by Dean Radin, currently at the Institute of Noetic Sciences, as well as by Dick Beirman, a professor at a university in Germany, have separately and collaboratively documented many examples of group consciousness effects on REGs, even over thousands of miles of distance. I know that this sounds hard to believe, but when we let the data speak, sometimes the data say very strange things; in quantum physics, there is actually a term for such things: they're called "quantum weirdness"!

My colleagues and I conducted an experiment that truly challenges the commonsense belief that randomness can exist in nature. We collected REG data twenty-four hours a day from Friday through Sunday over three consecutive weekends, with a pre-test weekend baseline in Arizona, an experimental weekend in New York, and a post-test weekend baseline in Florida.

The experimental weekend was at an international Qigong conference in New York City, where approximately twenty-five hundred practitioners of Yan Xin Qigong, a particular form of Chinese meditation, met for a conference. I brought a laptop computer and the REG device to the meetings, and recorded the REG data both in the meetings and during the evenings back in the hotel. The meetings included group meditation practices, watching dancing, hearing meditative music, and listening to Qigong lectures.

There were three possible predictions.

Prediction 1: There would be no REG organizing effects during the conference periods. The conference periods would look like non-conference periods recorded as baselines in Tucson and Boca Raton.

Prediction 2: There would be REG organizing effects observed during the conference periods, but the effects would disappear during the night and also during periods when the conference broke up into groups. This was the prediction my Western U.S. collaborators made.

Prediction 3: There would be REG organizing effects observed during the conference periods, and these effects would carry over and continue, even when the device was moved back to the hotel. This was the prediction made by my Eastern Chinese collaborators. They claimed that the group *qi* continues even during the evenings, since they meditate and interact virtually nonstop for the entire weekend.

The Evidence: What did the data say? The Western scientists were wrong: the data supported Prediction 3. Much to my surprise, the REG continued to deviate from chance throughout the entire weekend. Was this just a New York City effect? Or did this have something to do with the Qigong conference? (The questions are tantalizing, and I look forward to conducting follow-up studies.)

However, whether the effect was due to group Qigong, group process per se, New York City group consciousness, or other possible sources of synchrony is not important here. What's important is that reliable deviations in REGs could be observed, even in group contexts, and that the behavior of electrons in electromagnetically shielded devices is not immune from the organizing influence of the human mind—individually or collectively.

Mind may be able to organize matter in ways that the mind has yet to even imagine. As Paul Pearsall describes in *Wishing Well,* and Larry Dossey in *Reinventing Medicine,* the mind is more than amazing—it may be an expression of the ultimate organizing process, a Universal Organizing Consciousness.

In a most remarkable series of recent studies, William Tiller, professor emeritus from Stanford University, and his colleagues have been documenting how mental intentions can be imprinted in an electronic device that then alters the structure of systems, even when the devices are shipped and tested thousands of miles from where they were imprinted. In my favorite example, Tiller and colleagues have the devices imprinted with the intention that the pH of water will be increased. They then ship the devices, along with control devices not imprinted with intention, across the country for blind testing in a laboratory. The imprinted devices, when placed near water, produce an alteration in the pH of water as recorded by a computer. The control devices do not.

Tiller proposes that the mind is the original and ultimate organizing process. This parallels what the physicist Gerald L. Schroeder, Ph.D., author of *The Science of God,* proposes in the quote that introduces this chapter.

These are indeed paradigm-shaking experiments whose implications change our visions of everything. The general predictions follow directly from the integration of contemporary physics with systems science, and they point to the existence of a G.O.D. process in the universe as a whole.

IS THE AGE OF PERCEIVED INDEPENDENCE COMING TO AN END?—IMPLICATIONS OF UNIFIED FIELD THEORY FOR THE INTEGRATION OF SCIENCE AND SPIRITUALITY

The age of independence may be coming to an end. As Einstein said it, "Our experience of separation may be an illusion of consciousness."

The new science of dynamic interconnectedness, as expressed in Ervin Laszlo's *The Whispering Pond: A Personal Guide to the Emerging Vision of Science,* leads us slowly but surely to an alternative vision of a creative unfolding, Guiding-Organizing-Designing universe, and therefore the end of randomness as we once thought we knew it.

In the process, this emerging vision provides us with "a reason to hope" (the title of R. Wayne Kraft's visionary book). Everything may ultimately matter, including our wishes and even our prayers.

The great question becomes the origin of intention in all things everywhere.

Even what we see as disease may be revised by the emerging vision. Here is what I wrote in my 1983 master lecture to the American Psychological Association; the "third perspective" referred to is a special kind of complex order that in layman's terms we think of as *dis*order.

The implications of this third perspective for viewing health suggests that what are typically called diseases are not random disorders. Rather it is proposed that disease serves some ordering function for the evolution of the human species and nature in general. This ordering perspective encourages researchers to attempt to create a "periodic table of diseases" just as others have created a periodic table of the elements. . . .

This analysis implies that there may well be a wisdom [intelligent evolution] to disease, and that people may need disease in order to help them be healthy. In other words, the existence of disease provides people with essential feedback that informs them of the fundamental rules for survival and evolution. If they listen to this feedback, they have the potential to evolve to the extent that they recognize how these fundamental limits and regulations are essential for survival and growth.

However, it is now the twenty-first century, and biology can guide us only so far. Our capacity to evolve our mind and consciousness is key, in addition to academic knowledge. With knowledge comes the need for the emergence of wisdom and taking personal responsibility.

It may be time that we start taking personal responsibility for the organizing potential of our minds, and learn how to become a wise species. Though it has been wise historically to separate Church (religion) and State (government), it may be time scientifically to connect G.O.D. (universal spirituality) with State; for example, "In G.O.D. We Trust."

Imagine that you have stepped outside the
whole universe,
And you are seeing it as a Gigantic Caterpil-
lar of Living Galaxies,
Slowly but surely readying to transform
into a
Rainbow Colored Butterfly universe of inde-
scribable beauty . . .

SAM

16

Wisdom in the Stars

IS THE UNIVERSE NOT ONLY INTELLIGENT
BUT WISE?

Many women and men throughout recorded history have dreamed of the possibility of homo sapiens evolving into something grander. Evidence of this potential evolution can be witnessed by examining our species history with integrity. If we focus on the positive, we can see how religion, science, technology, politics, education, medicine, law, business, and so forth have all evolved into ever more complex systems *capable of great power and goodness.*

For example, as Dr. Jonas Salk proposed in his prophetic book *The Anatomy of Reality: Merging of Intuition and Reason,* evolution, especially human evolution, has progressed from being survival of the fittest through survival of the smartest to survival of the wisest. In a word, Salk is proposing that evolution is evolving. Yes, evolution itself is evolving.

In medicine, there was a time when surgery was conducted with nonsterilized, relatively dull knives on patients who were awake. Surgery has evolved procedurally and technologically. The tools are becoming ever smaller and gentler—knives can now be directed by fiber optics, even by computers, and other forms of "knives" such as tiny lasers are becoming available. Anesthetics have evolved as well, and through contemporary versions of ancient techniques such as electroacupuncture and guided-imagery hypnosis, pain is being lessened. Clearly, our bodies must be fit for us to survive biophysically. And visionary individuals like Dr. Andrew Weil, on the faculty at the University of Arizona, are leading that parade of integrative medicine toward wellness.

But the time has come for people to be smart and intelligent in addition to being fit, healthy, and strong in order to survive. And it is becoming clear that our planet will not survive, and therefore we won't either, if we do not learn how to use the evolving knowledge wisely— and that means with courage, integrity, caution, responsibility, kindness, and humility. Humans have the capability to become an evolving wisdom species that can transform itself beyond its caterpillar past history and become its butterfly future.

It appears that as a species we have the intellectual potential not only to destroy or disintegrate everything (our destructive side), but to integrate everything (our constructive side) as well. If all things are indeed interconnected and interdependent, then we are all part of the Almighty One. In fact, according to contemporary field physics, we are never alone—rather we are "all-one." We need to learn to experience atonement, which is "at-one-ment."

Can we envision our inherent evolutionary potential? Is it time to change direction and see with new eyes? Is the answer simply more technology, or must we see the universe in an alternative or new way?

The Buddha said: "If we do not change directions we will end up where we are going."

As a young assistant professor of psychology and social relations at Harvard University, more than thirty years ago, I accidentally followed a train of logic that frankly terrified me. I ultimately published a scientific description of the logic as a scholarly chapter in an academic book in 1977, but my colleagues in psychology and medicine were clearly not interested in the vision.

The logic, though compelling, was too unifying and covered too many disciplines and levels of nature to hold their attention. In addition, the argument would force all who followed it to reevaluate the underlying logic of doing what we were doing (and going where we were going). At that point in time most of my colleagues were opposed to any professional and personal soul-searching. They needed to focus on their careers and success. My vision was distracting and therefore not to be considered.

The vision came to me in a way that was both humorous and humbling.

THE LESSON OF THE SUPERSTOMACH—HOW EVERYTHING HAS THE POTENTIAL TO GO OUT OF CONTROL

In the spring of 1972, I had recently completed my Ph.D. at Harvard, and I was teaching my first large undergraduate lecture course. It was a course that I had never taught before, and to the best of my knowledge, it had never been taught before by anyone else. This new course bridged biology, personality, consciousness, and health. It had the unusual title Biopersonality and the Mind of Man.

Since there was still carryover from the sixties, the topic attracted hundreds of students, who were to attend a single two-hour lecture on Wednesday afternoons along with weekly class discussions in small groups that were taught by my team of graduate teaching fellows. Since there were no existing textbooks, I pulled together diverse references from many sources to comprise the unique reading list.

I began each week with a rising anxiety as I had to read the upcoming materials to prepare for the two-hour lecture. Like clockwork, Wednesday morning was a superstressful time for me, and I regularly had a stomachache. I would spend significant time reading and organizing the materials and trying to quiet the discomfort of my stomach.

The pain was severe at times, but I could see no way out. I had created this unusual course in the first place, and I was determined to give the students the best lecture possible. I had promised a lot and wanted to deliver even more.

Three quarters of the way into the semester, on one of those Wednesday mornings, I was preparing to give a lecture on biofeedback and psychosomatic diseases that would address the interrelated topics of stress, feedback, self-regulation, and health. My body was fighting me and I was hurting and feeling sorry for myself. I wondered why I, with the iron stomach, had to have this digestive ache.

For the afternoon lecture I was preparing to talk about Dr. Walter Cannon's book *The Wisdom of the Body.* So I paused to examine the concept of pain as being a feedback process between my body, brain, and mind and I asked myself the question, Is my pain serving a purpose? And I immediately realized that my digestive pain was telling me that I was living my life in a manner that was too stressful and unhealthy. My stomach growled at me in a private language that I not only could hear, but could listen to. The pain was trying to slow me down, to focus my attention on what I ate and how I was living. My body was telling me that it desperately needed help, and that I must cooperate. Pain was pointing me in a better direction, whether I wanted to listen or not.

However, what came to me was that our society, then, was not teaching us that pain was actually friendly feedback helping us to choose to live healthier and wiser lives. Quite the contrary, the commercials on television were blasting a different message, and even scientists like me were buying the message that pain was my enemy. For example, my mind leapt to the award-winning Alka-Seltzer commercials of the time. One showed grossly obese men at a pie eating contest, stuffing their faces with apple pies to win a prize. In the middle of the contest, one of the men got a stomachache. The message of the commercial was not "Stop eating too much food quickly." The advertisers were reassuring as they said, "Eat! Eat! Don't listen to your stomach— pig out and take Alka-Seltzer afterward."

Well, what about the wisdom of the body? This was a case of feedback, particularly what is termed in systems theory "negative" feed-back (which in this case means subtraction). Negative feedback causes the system to change directions mathematically so as to maintain its stability. For example, when a room gets cold, the furnace turns on, because the thermostat is connected as a negative feedback device, sig-naling the furnace to create heat and return the room to the selected temperature. While feedback serves a fundamental self-regulating function, negative feedback helps control the behavior of a system and changes its direction. Imagine what would happen if the pain negative feedback loop—this "change of direction" feedback loop—was artifi-cially disconnected?

I called Stephen Krietzman, Ph.D., the person who taught me the black box story when we were both graduate students. I had been amusing myself to help withstand the pain by indulging myself with design ideas for a high-tech superstomach that would never experience pain, would not be subject to obesity, and would not require conven-tional food, so it would not contribute to depleting the food resources of planet earth.

As an expert in nutritional biochemistry, and a friend who under-stood my sense of humor, Stephen said, "Gary, no problem. We can design your superstomach so that people can eat the leaves on the trees. The reason why humans cannot eat leaves is because their stomachs cannot digest them. What your superstomach can do is make it possi-ble for people to digest leaves. Then, through the evolving miracle of modern nutritional biochemistry, we can learn how to cook with leaves to make them tasty—even taste like meat and vegetables."

I was delighted with Stephen's input. Not only would my fantasy superstomach eliminate pain and solve obesity and make me my first million, it would eliminate world hunger, too! But the humorous ad-ventures of my creative brain did not stop the pain I was suffering.

I had still to learn the implicit universal lesson of pain—my cre-ativity and active mind were taking me further away rather than help-ing me to understand the fundamental message I was getting from the feedback called pain. The fact is, the whole earth could potentially go out of balance if the stomach feedback loops are disconnected. As we eat more, we deplete more; even carbon dioxide could go out of control as the remaining leaves on the trees were consumed by humans.

Norbert Weiner—the great mathematician at MIT who happened to be a postdoctoral fellow of Cannon—discovered that no system can exist without feedback. Feedback is universal. Feedback appears to be the foundation of life, memory, and evolution. Feedback appears to be what enables everything to communicate, share, connect, and grow. Feedback is the greatest gift any system has been given. No feedback, no survival. No feedback, no intelligent evolution. Feedback is the $E=mc^2$ of survival and evolution. It's that simple.

The fault is not with our technology per se. Nor is the fault in our temperament. The fault is in our relative ignorance in understanding how systems work, live, thrive, and evolve. People are slowly learning the ground rules of the game of life, and some of these rules are neither easy nor likable. However, these essential rules, apparently established by an intelligent G.O.D. process, reflect the fundamental scaffolds required for both existence and evolution. If we can come to understand and honor these protective and organizing feedback fields of guidance, our potential to unfold should be enhanced.

It's worth remembering Pagels's comment about how randomness is presumed to occur only in the absence of rules. Weiner's universal feedback (or "field-back") may be the rule of rules. Appreciating this grand rule is now within our grasp.

ROLE MODELS OF EMERGING HUMAN "BUTTERFLIES" THAT INTEGRATE SCIENCE AND SPIRITUALITY

History is replete with potential role models who have envisioned a grand organizing universe and the great potential for humanity within it.

These women and men have come from all walks of life, including science and religion. In this book, I can only briefly acknowledge the existence of their visions. There have been many human butterflies that point the way to our capacity to evolve and soar.

Since this is an evidence-based faith book, I wish to mention two scientists—Rustum Roy, Ph.D., and William Tiller, Ph.D., both of whom I have come to know personally, whose academic careers and writings point to the remarkable potential for human evolution. Roy has spent his career primarily as a distinguished professor at Penn State University, not only publishing cutting-edge science in mainstream journals, but also attempting to inspire science to build bridges within its own community, and even more between its own community and the community of religion.

Roy's book *Experimenting with Truth: The Fusion of Religion with Technology Needed for Humanity's Survival,* published in 1981, based on the distinguished Hibbert Lectures he gave in 1979, was clearly ahead of its time. I believe this book should be required reading for anyone interested in the evolution of science and religion as they move toward a common synthesis and understanding. Roy presents readers with an evolving vision of reality; as he explains, his book "has explored the areas where the insights of science and religion may feed each other for the benefit of humanity." It's time for the mutual feeding of these two great paths of human evolution.

Similarly, Tiller has spent his academic career as a pioneering professor at Stanford University, not only publishing in mainstream scientific journals, but also attempting to inspire science to build bridges between itself and spirituality. In his book *Science and Human Transformation: Subtle Energies, Intentionality, and Consciousness,* published in 1997, Tiller reviews his controversial research in frontier physics that challenges us to think beyond today's accepted theory to a new vision of physics and consciousness. Though not easy reading because it requires processing mathematics and information from many disciplines, the science chapters are mixed with chapters like "Inner Self-Management," "Training Wheels for Humanity," and "Stewardship of the Earth" that all of us can understand.

Of the two, Roy is the more conservative, remaining in the mainstream and becoming a member of the National Academy of Engineering. Tiller made the decision to address the most controversial topic of all—the physics of spiritual reality—not only theoretically but experimentally. Both have received criticism from some of their peers for going "too far too fast."

The question arises, have these two visionary scientists lost their minds, as their harshest critics suggest? Or have they actually found their souls? The evidence strongly points to the latter as being the correct interpretation. I suggest that metaphorically Roy and Tiller are examples of scientist butterflies, illustrating what is possible for all of us. As Tiller recently said in a lecture I attended, "If one person can do this, then everyone can."

EVOLVING MEANINGS OF GOD, SPIRIT, AND SOUL

Terms like "God," "spirit," and "soul" can evolve in meaning as our understanding of reality evolves. Science and religion will come together when they are each willing to accept the fact that history is a process of learning and growing, and that knowledge is continuing to unfold. Our past stories are just that—stories—and they should evolve as our knowledge does.

Can we envision spirit and soul as wearing contemporary scientific clothing? I wrote the following poem, revised slightly for this book, to explore this potentially unifying vision.

SOUL AS INFORMATION, SPIRIT AS ENERGY: MANIFESTING THE EVOLVING FIELDS OF WISDOM AND LOVE

What, pray tell, are Spirit and Soul?
Are they one and the same?
Are Soul and Spirit a functional Whole,
Derived from a common Name?

Or is it the case that Soul and Spirit
Reflect two sides of a coin,
Where Soul expresses the Information that fits,
And Spirit, the Energy that joins?

Is Soul the story, the Plan of Life,
The music we play, our score?

Is Spirit the passion, the Fire of Life?
Our motive to learn, to soar?

Soul directs the paths we take,
The guidance that structures our flow.
Spirit feels very alive, awake,
The force that moves us to grow.

If Soul is Plan and Spirit is Flame,
Then fields are alive, you see.
Nature may play a majestic game,
Of Information and Energy.

I'd love to believe that wisdom and joy
Reflect God's plans and dreams,
That Soul and Spirit are more than toys,
And both are more than they seem.

Could it be that the Soul of God
Is the wisest of plans, so grand,
And the Spirit of God is the lightning rod
That inspires God's loving hand?

Could Soul be Wisdom, and Spirit be Love,
Together, a divine partnership,
Purpose and passion, a Duet from Above,
The ultimate relationship?

The relationship of Spirit to Soul,
So simple, profound this team.
For Spirit and Soul the ultimate goal,
To understand this scheme.

Soul as Wisdom, Spirit as Love—
Information and Energy,
Awakened compassion, the Flight of the Dove,
Someday, pray tell, we'll see.

Are human beings a core expression of this invisible organizing G.O.D. process? Do we have the potential to discover scientifically the existence of an invisible organizing reality, a universal organizing consciousness? And can we evolve with it? Can we become *homo spiritualis*?

This book concludes "Yes." The story, and the evidence, must continue.

When you close your doors, and make darkness within,
remember never to say you are alone, for you are not alone;
God is within, your genius is within.

EPICTETUS

17

The Genius Within Everyone

WHY UNIVERSAL ORGANIZING CONSCIOUSNESS EXISTS IN EVERYTHING

In this book, in which G.O.D. is a universal Guiding-Organizing-Designing process, I am also implicitly saying that G.O.D. can be thought of as some sort of omnipresent, superintelligent "genius-optimizing developer" of the universe and everything within it. Given that science is leading us to some sort of a genius G.O.D. process, can science take us further and reveal more precisely what kind of G.O.D. process is God? To that question I also answer yes.

But I use the term "believe" in the same sense that, as mentioned earlier, the related word is used in the last line of the movie *Dragonfly:* "Belief is what gets us there." And this meaning is important.

Again, it is not enough to know that crafting a sand painting and building a skyscraper are possible; *you must believe that you can do them.* It is not enough to know how to do these things; *you must have the motivation to achieve them.*

In a deep and accurate sense, belief is what ultimately leads people to create and to evolve. It is in this sense that I believe in science. In this book you've learned that I have come to believe science can provide definitive answers to fundamental questions about the nature of reality. My belief in the power of science is evidence-based. The history of science contains abundant evidence of its potential for discovery, understanding, and application. And because I believe in science, I am motivated to apply the methods of science to questions that are seminal, regardless of how controversial they may seem.

You've discovered in these pages my efforts to create the optimal experimental conditions for the chance explanation—the conventional, presumably correct explanation for the origin and evolution of the universe and its laws—to prove itself. My work has given the chance explanation every chance to prove itself. The bottom line is that when chance is given the chance to prove its assumed fundamental role in the creation of order in the universe, it ends up disqualifying itself.

My belief is that if chance were the correct answer, it would be revealed in experiments, and if chance were false, we would find the mistake in our interpretation. The fundamental mistake is explained in Chapter 10—that order does not occur by chance (it requires precise conditions); instead, chance gives the opportunity for creative orders to occur. So, returning to the G.O.D. question and the genius within, can science slowly but surely reveal the precise nature of the G.O.D. process? Can science someday answer Einstein's question about the detailed "thoughts of God"? Again, based upon already existing evidence combined with straightforward reasoning, my answer is a qualified yes.

Cognitive neuroscience and computer science have made tremendous progress in exploring and understanding the nature of intelligence in both natural and man-made systems. Natural intelligence and artificial intelligence—the latter, of course, implemented by human intelligence—are both evolving at breathtaking speed. The process suggests to me the way the individual atoms of hydrogen and oxygen come together, and are modulated by outside fields (such as information and energy conveyed at room temperature—recall Chapter 11) to make the emergent liquid molecule called water, which under the right conditions can become snowflakes of many shapes.

In the same way, individual components (be they physical, biological, social, or informational, as in computer software code) can come together, be modulated by outside fields, and make emergent feedback systems termed networks—with uniquely new emergent properties that are inherently intelligent. The more complex the feedback network and its interconnections, the more complex and unique are the intelligence and the associated learning capacities.

Amazingly, we now take the existence of man-made intelligent feedback network systems for granted. The level of intelligence programmed into our contemporary cell phones is extraordinary and is growing in each generation of digital phones. These evolving intelligence-driven cell phones interconnect us and thus create ever more complex feedback loops between many of us on the planet. Intelligence begets intelligence. Feedback loops are a universal phenomenon at every level of nature, and their expression is evolving via intelligence.

One might wonder if consciousness is a universal phenomenon like feedback. And then we can wonder: Does consciousness exist only in humans? Does consciousness extend to dolphins, dogs, and African grey parrots? Does consciousness extend to butterflies, protozoa, and *E. coli* bacteria? Does consciousness extend to trees, cacti, and pollen? Does consciousness extend to mountains, rocks, and raindrops? Does consciousness extend to atoms of hydrogen and oxygen, or even massless photons that are infinitely small? Does consciousness extend to the earth, moon, and sun? Does consciousness extend to solar systems, galaxies, and superclusters of galaxies? Does consciousness extend to the universe and beyond? And does consciousness exist in all feedback network systems at every level of nature and the universe?

You must ask how you personally draw the line for consciousness. Is your opinion based upon your education and culture, emotion and wishes, personal experience, logic and reasoning, or experimental evidence? Philosophers hold vastly different opinions on consciousness, ranging from "There is no such thing as consciousness" to "Consciousness is the fundamental building block of everything in the universe, from nonphysical intention and information to physical energy and matter." On the other hand, a nearly unanimous opinion is held by the founders of the great religions that an omnipresent and omniscient God has an infinite mind with Universal Consciousness.

Throughout recorded history, philosophers have viewed God as purportedly "Conscious" with a capital *C*. Their God is Consciousness. If God is Consciousness, and if God has omni-awareness, then everything that exists in the universe has God or divine consciousness within it. Because if "the All is in the Small" and the G.O.D. process (that is, the All) equals consciousness, then consciousness is in the Small and everything has consciousness expressed to various degrees.

If consciousness is the foundation of all that there is, then part of our awakening as an evolving species is our capacity to discover this potential fundamental truth about nature and the universe. However, the question must be asked if science can determine whether this logic actually matches reality. Does any existing scientific evidence lead to the conclusion that the G.O.D. process reflects a universal organizing consciousness whose omniscient genius is beyond anything we can imagine?

APPRECIATING ANOMALIES IN SCIENCE

A growing body of scientific evidence, dismissed by many contemporary scientists as being anomalous if not impossible, is slowly but surely emerging to support such a G.O.D. presence. Because this body of scientific evidence inherently integrates science and spirituality, it is resisted if not rejected by many authorities in our contemporary educational and religious institutions. Fortunately, this era of denial and rejection is coming to an end.

Professor William James, M.D., deeply appreciated the importance of anomalous and "irregular" phenomena that may seem to be impossible and "unimaginable." He wrote at Harvard almost a century ago in the English of his era:

> *"The great field for new discoveries," said a scientific friend to me the other day, "is always the unclassified residuum." By that curious term, he meant that around the accredited and orderly facts of every science,*

*there ever floats a sort of dust-cloud of exceptional observations, of oc-
currences minute and irregular and seldom met with, which it always
proves more easy to ignore than to attend to.*

The passage goes on to argue that "each one of our various *-ologies*
seems to offer a definite head of classification for every possible phenomenon of the sort which it professes to cover." And he found most
men "so far from free . . . that, when a consistent and organized
scheme . . . has once been comprehended and assimilated, a different
scheme is unimaginable. No alternative . . . can any longer be conceived as possible." Anything that didn't fit would be seen as "para-
doxical absurdities, and must be held untrue."

When exceptions are demonstrated, they are "neglected or denied
with the best of scientific consciences." Only the born geniuses, James
wrote, would become "worried and fascinated by these outstanding
exceptions, and get no peace until they are brought within the fold."
And he pointed to scientists like Galileo and Darwin, who, he maintained, were "always confounded and troubled by insignificant
things."

Instead of being stuck with old ideas and unwilling to accept anything new, James wanted to see a scientist be able to "renovate his science." He would want *all* of us to be willing to renovate our ideas, as
well.

EVIDENCE THAT POINTS TO A CONSCIOUS UNIVERSE

Dean Radin, Ph.D., wrote a pathbreaking book titled *The Conscious
Universe.* This book provides a readable and careful review of a large
body of scientific research in what is often termed parapsychology.
Radin reviewed hundreds of carefully conducted experiments documenting the existence of what others have called the *extended mind* or
the *nonlocal mind,* and Radin calls the *entangled mind.* This research
not only demonstrates the capacity of one person to sense the thoughts
and feelings of others—even if they are separated by hundreds or
thousands of miles—but it documents the capacity of one person's

mind to influence the behavior of quantum events such as electrons in electronically and magnetically shielded devices.

Research my colleagues and I have conducted on the organizing mind, presented in Chapter 15, replicates and extends this body of knowledge in an important manner. It is possible to demonstrate that either individual minds, or groups of minds working together, can have unintentional and unconscious effects on the quantum behavior of electromagnetically shielded electrons. In a word, our minds can organize the physical world at the quantum level. Now, if consciousness is indeed universal and omnipresent (in other words, is not limited to humans), and is primary (in other words, universally creates information, energy, and matter), it follows that everything that exists in the universe may ultimately be interconnected and organized via consciousness.

Certain states of consciousness may invite resonance if not unity between individual systems in the same way that musicians can play together in an orchestra or singers sing together in a choir. When people enter different states of consciousness, trillions of electrons in our brains synchronize in definable ways; we can measure this synchrony as brain (EEG) waves or fields.

Apparently, electrons that are electromagnetically isolated in a shielded device can nonetheless resonate with our brain's synchronized electrons under certain conditions of consciousness. But does consciousness itself show emergent network properties? Does consciousness show novel properties that emerge from individual electrons to individual brains to groups of brains to a global brain? Is there such a thing as emerging global consciousness? Does the earth have consciousness, and are we—both individually and collectively—components of this consciousness?

Roger Nelson, Ph.D., and colleagues from the Department of Engineering at Princeton have been conducting research termed the Global Consciousness Project (see http://noosphere.princeton.edu). Here is how Nelson et al. describe the nature of global consciousness:

Research on human consciousness suggests that we may have direct communication links with each other, and that our intentions can have effects in the world despite physical barriers and separations. We are compelled by good evidence to accept correlations that we cannot yet explain.

Princeton's Global Consciousness Project speculates that fields generated by individual consciousness would interact and combine, and ultimately have a global presence. Occasionally, the Nelson team explains, "global-scale events . . . bring great numbers of us to a common focus and an unusual coherence of thought and feeling." To study the effects of a possible global consciousness, the Princeton group has "created a world-spanning network of devices" and believes the data they have collected so far appear to correlate with events "that may evoke a world-wide consciousness." The events they list "include both peaceful gatherings and disasters: a few minutes around midnight on any New Year's Eve, the first hour of NATO bombing in Yugoslavia, the Papal visit to Israel, a variety of global meditations, several major earthquakes, and September 11, 2001."

When science combines this evolving body of experimental research with contemporary laboratory research on the topic of survival of consciousness after physical death, we discover compelling evidence that leads to a notable conclusion: that the brain is not required for conscious experience, intention, and intelligence (recall our discussion of the mind-brain relationship question in Chapter 11). Furthermore, as summarized in the last chapter of *The Afterlife Experiments,* the totality of the evidence leads inexorably to the conclusion that the answer to the question "Which comes first, brain or mind?" is "Mind first, then brain." Brain becomes a physical tool of the mind, not the other way around.

William James was a proponent of the "mind first" thesis. Probably the most forceful contemporary proponents of this position are Stanford professor emeritus William Tiller and his colleagues. In their book *Conscious Acts of Creation,* they begin by saying: "This book marks a sharp dividing line between old ways of scientific thought and old experimental protocols, wherein human qualities of consciousness, intention, emotion, mind and spirit cannot significantly affect physical reality, and a new paradigm wherein they can robustly do so!"

I admire the use of that term "robustly." Their book gives evidence not only for the organizing power of the human mind, but for the intimate interconnectedness of all things via universal organizing consciousness. And support of this concept is assembling from many sources.

FROM THE CONSCIOUS UNIVERSE TO INTELLIGENT EVOLUTION AND G.O.D., THE ULTIMATE EXPERIMENTER

Most scientific researchers never mention the G-word in the context of their science. Unfortunately it is even more taboo in the research laboratory than it is at most dinner parties. Nor is the G-word listed in the index of books like *Conscious Acts of Creation* or *The Field,* or mentioned, except briefly in passing, in books like *The Genius Within* or *The Conscious Universe.* However, the avoidance of the G-word reflects the *politics* of science, not its essence. For when the emerging body of consciousness research is "seen with new eyes," it portends a spectacular paradigm change.

Professor Rustum Roy describes the work of Tiller and his colleagues as follows: "For the first time, in the language of physics, very solid and very extensive data on 'Spirit>Mind>Physical Matter' interactions have been provided." Science and spirituality are becoming two sides of the same coin. Science is not only helping to solve the mystery of mysteries, it is serving the mystery of mysteries. In fact, we are beginning to see the G.O.D. process as the Ultimate Experimenter, an intelligent, creative, and caring intellect that we can come to know and serve—if we are willing to ask.

QUESTIONS UNANSWERED

My evidence-based faith is that if we stay willing to continue asking challenging questions with an open and discerning mind, and then we creatively apply the evolving tools of science to address these questions, definitive answers will be revealed in time. It is the nature of science that when one question is answered, more questions present themselves.

And so, as the end of this book approaches, you are left with many unanswered questions, such as: Is what we term the G.O.D. process an expression of a Universal Organizing Consciousness—an awareness and intelligence in everything? Is the G.O.D. process perhaps conducting a great experiment? Is the ultimate outcome of this grand experiment already known (and is it possible that the G.O.D. process has a bag of cosmic tricks up the sleeve)? Or is G.O.D., the Ultimate Experimenter, continually discovering new things just as we are? And is the G.O.D. process itself evolving along with the evolving universe?

And a question we all share: *Is there a universal co-creative process of which we, individually and collectively, are a part?*

Hopefully we'll receive the answer. I am sure it will be provided if we become mature enough and wise enough to apply our knowledge and science in a way that is even more compassionate, healing, and transformative. I personally am committed to continuing this work for our species and our precious planet. The future is in the hands of all of us.

> *The intuitive mind is a sacred gift and the rational mind is a faithful servant. We have created a society that honors the servant and has forgotten the gift.*
>
> ALBERT EINSTEIN

18

Summing up the G.O.D. Experiments—The Emerging Case for Intelligent Evolution

SEEING THE BIG PICTURE, AND NEVER FORGETTING IT

Einstein calls on us to use not just our rational mind but our intuition. As we review the evidence presented in these pages, that's a worthwhile admonition to keep in mind: remember not only to honor your rational mind as a "faithful servant" but to also envision your intuitive mind as a "sacred gift" from the Source—even if you see the Source simply as your own mind. Tap into your intuitive side as well as your rational side.

THE BIG QUESTION

Is there something fundamental that needs to be explained in nature and the universe? The answer is clearly yes. The far-reaching question has to do with the overabundant evidence of evolving orders in the universe. Everywhere we look, from photons and chemicals, through flowers and people, to galaxies and superclusters, we witness dynamically unfolding orders. And the majority of these evolving orders are indescribably complex as well as breathtakingly beautiful.

Order appears to be the rule, and not the exception, in nature. Some of the orders have been termed by Tyler Volk, Ph.D., "metapatterns"; these replicate themselves at every level in the universe. They are *universal*.

One of the most ubiquitous metapatterns in nature—a pattern that follows the "Golden Ratio" called phi—is the universal spiral shape that appears in the trails made by subatomic particles, in the double-helix molecule DNA, in seashells, in the movement pattern of the heart, in tornadoes, in galaxies.

Over the past two centuries science has catalogued a dizzying array of exquisite orders. Scientists have created optical microscopes and telescopes that enable researchers to witness orders that are too small or are too far away to be seen unaided. Electrical engineers have created supersensitive sensors that measure electromagnetic fields vibrating at frequencies other than the restricted band of frequencies visible by the human eye. Researchers can now examine a myriad of invisible orders—using microwaves, infrared waves, ultraviolet waves, radio waves, X rays, gamma rays—manifesting from subatomic particles to superclusters of galaxies.

It is also possible to witness the evolution of complex orders that are created by humans. Scientists of the twenty-first century function in a veritable sea of intelligently guided experimental designs expressed in architecture, art, music, science, and technology. The evidence from history convincingly demonstrates that it is possible for us to function brilliantly in the human role of being an extraordinary Guiding-Organizing-Designing species.

Science can be described as a formal process of discovering—through intelligent trial-and-error experimentation—the existence of order where previously we thought there was no order. Science discovers new patterns and then attempts to explain them.

The question is, does the ubiquitous existence of these evolving orders reflect the expression of a random evolutionary process, or an intelligent evolutionary one?

ORCHESTRATED ORDERS IN OUR PERSONAL LIVES

How wide and how deep does order extend in the universe? Does it extend all the way to the orchestration of our daily lives? If order is the rule in the universe, and not the exception, does complex ordering apply to our personal lives as well? You have already encountered in these pages a few detailed examples of extraordinary synchronicities in human life that illustrate the kind of amazing evidence that can be obtained on this profound question.

The unanticipated findings from the Chris Robinson experiment illustrate how contemporary precognitive parapsychology research can uncover uncanny patterns in the coordination and sequencing of our daily lives that seem utterly incomprehensible to most of us. The unanticipated synchronicities or coincidences revealed in the "Remember the diamond" story provide compelling evidence of everyday orchestration that can be observed not only in controlled parapsychology experiments (Part Two) but in our personal lives as well (Part Three). And a truly exceptional avalanche of synchronicities is revealed in Appendix C.

However, we will never discover such synchronicities in the laboratory or in our lives unless our minds are open to seeing them and we are willing to systematically record the observations. This is a case where belief in the possibility of order is a prerequisite for seeing evidence of order in science as well as daily life.

THE UBIQUITOUS NATURE OF ORDER

If we accept the evidence not only from physics, chemistry, biology, ecology, and astrophysics, but from controlled parapsychology experiments as well as personal exploratory experiments, that order is the rule, not the exception, in nature, then how do we explain it? Where does all this order come from?

As we have discussed, the simplest explanation is to suggest that orders can occur and evolve by chance alone. Sometimes called blind chance, this is the generally accepted explanation from conventional science. It is generally assumed that randomness allows for all possible orders to occur, and by chance, evolution—including natural selection—will unfold.

It is further generally assumed that the universe is becoming more disordered over time, as described by the Second Law of Thermodynamics. Science generally assumes that the emergence and evolution of pockets of order existing in the universal sea of increasing disorder are due to the operation of chance combined with what is termed "self-organization." Self-organization is presumed to be the capacity of certain systems to self-assemble without the aid of an external organizing force. Darwin's natural selection explanation combines chance mutation with self-organization. But is the blind chance explanation necessarily the correct explanation? No, not necessarily.

The alternative explanation involves some sort of creative intelligent, trial-and-error ordering process—what I term an Experimenting Guiding-Organizing-Designing (G.O.D.) process. This explanation proposes that what Darwin termed "natural selection" is an intelligently guided experimenting process—the G.O.D. process—that promotes the creative expression and evolution of ever more complex and interconnected systems. What Darwin termed "natural selection" may actually reflect the operation of "intelligent selection."

Why would I propose that people consider entertaining an alternative hypothesis to the blind chance explanation? Remember, the question being asked is: can scientific evidence combined with careful reasoning lead us to conclude that some sort of an Experimenting G.O.D. process exists in the evolving universe?

There are three primary reasons for seriously questioning whether the conventional, generally accepted chance explanation is correct. As noted earlier, Reasons 2 and 3 are provided for the first time in this book.

REASON 1: THERE IS TOO MUCH PRECISE ORDER IN THE UNIVERSE TO HAVE OCCURRED BY CHANCE.

The combined evidence spanning physics and chemistry to ecology and astrophysics suggests that there is too much order in the universe. Recall how microscopes and telescopes reveal a previously unseen wealth of order. This wealth of order is organized in too precise a manner to plausibly be explained as occurring by chance. The balance of forces necessary to sustain the universe and ultimately sustain biological life is too precarious and precise to have come into existence simply by chance alone. The evidence and the reasoning for this position are provided in a number of recent books, such as *The Anthropic Principle*. The interested reader is encouraged to review this emerging abundance of evidence. (Recommended readings are included at the end of this book.)

REASON 2: THE CHANCE EXPLANATION IS NOT SUPPORTED BY EXPERIMENTAL EVIDENCE.

When experiments are conducted that actually test predictions made by the chance hypothesis, compelling evidence is discovered to indicate that the chance explanation does not work in practice. The simple physics experiments presented in Part Two of this book—focusing on a simple sand painting experiment—convincingly demonstrate that chance by itself cannot explain all of the order that is observed in the universe as well as in everyday life.

But the skeptic may wonder whether sand paintings (or clothes folding) might happen given enough time—say, billions of years. When you examine the data from the experiments carefully, you find no evidence whatsoever to support such a speculation. In the same way, watches, buildings, symphonies, computers do not come about by chance. These things require a sophisticated planning and implementing ordering process—that is, human designers and builders—to come into existence.

Such obvious simple physical observations are complemented by more complex but easily understood computer modeling experiments.

Recall that when conditions for producing random sampling are in place, in the absence of a specific ordering process, numbers always create what is termed a "normal" distribution, a bell-shaped curve. And the more numbers used in the calculations, the more perfectly formed becomes the bell-shaped curve.

The combination of evidence in Part Two from the simple physics experiments, plus the more complex computer modeling experiments, conclusively disproves the key prediction. Contrary to well-accepted popular belief, when we actually conduct the experiments, we discover that replicated organizations do not emerge by chance. It is sometimes said that only a donkey makes "ass-umptions." I too once ass-umed that the chance explanation must be true; I never tested whether it actually worked in the real world.

REASON 3: THE CONDITIONS FOR CREATING CHANCE DO NOT EXIST IN THE UNIVERSE.

Do the conditions necessary to produce random sampling exist in the universe? The answer to this fundamental question is *never,* and I emphasize the word "never." When we carefully consider the conditions necessary to produce random orders in the first place, we are reminded that a prerequisite for random sampling is independence of events.

If events are not independent, random distributions of events are never observed. Again: *never.* This is a well-accepted fact in science. Computer programs have been designed that intelligently model the creation of complex orders requiring the use of feedback loops. The loops in the program carefully connect the output of a given equation to the input of another equation. The equations are therefore interconnected; they are not independent. One equation feeds the other, and depending upon the program, this can happen hundreds of times or millions of times. Then, and only then, can replicable evolving orders be observed. So this does not meet the test for randomness. No ifs, ands, or buts.

Even what is termed "self-organization" only occurs when feedback loops exist and function in an organized fashion. Organization can create organization; however, randomization does not in and of itself create organization. So-called artificial-life programs require intelligence to program them and intelligent operating systems to carry out the instructions.

Does independence exist in real life? When we carefully look for the existence of independence in the universe, what we find instead is interdependence and interconnectedness. Contemporary physics tells us that everything that exists in the material universe is interconnected with everything, to various degrees, by invisible force fields of information and energy. These invisible physical fields include gravitation and electromagnetism. In other words, interconnection (like order) is the rule, not the exception, in nature.

What physics calls the "vacuum" or "void" is actually filled with incomprehensibly complex dynamic networks of highly organized—and organizing—fields of force. Though physicists do not understand the origin and organization of these invisible force fields, they are convinced that these invisible fields definitely exist and play a regulating or guiding role in all physical systems. Wireless communication by cell phones is one example that illustrates this. In the chapter titled "Can G.O.D. Play Dice with the Universe?" I discussed the typically ignored discrepancy between what statistics and physics tell us (randomness requires independence, but everything is in fact interdependent) and how we therefore erroneously apply the randomness explanation to the origin and evolution of order in the universe.

When we accept this fact, there is no longer a justifiable reason to expect that, given enough time, chance alone could create something as simple and beautiful as a Native American sand painting, let alone a single living cell, or a V-shaped formation of flying geese each composed of billions of organized cells.

In fact, the combination of evidence and logic leads to the paradigm-shifting conclusion that there is no such thing as pure randomness in the universe.

What we perceive as randomness may instead reflect a degree of complexity of order that we have yet to decode (and in some cases, as proved mathematically by Gödel—mentioned below—we may never be able to decode).

IF CHANCE IS NOT THE EXPLANATION FOR ORDER,
THEN WHAT IS?

Once you fully accept Reasons 1 through 3 above—that the existence of ubiquitous order in the universe (Reason 1) does not (Reason 2) and cannot (Reason 3) occur by chance—you are led inexorably to consider the possible existence of some sort of universal, invisible, intelligent Guiding-Organizing-Designing field process in the universe.

It follows that what I am calling an invisible Experimenting G.O.D.-field process (or simply the G.O.D. process) must somehow, to some degree, be playing a fundamental Guiding-Organizing-Designing role in everything that expresses order—from subatomic superstrings, through personal coincidences that are too unexpected to be by chance, to superclusters of galaxies.

Moreover, it follows that even when we observe apparently self-organizing systems—such as the formation of individual raindrops containing billions of atoms, or spiraling galaxies containing billions of stars—they too must involve, to various degrees, an invisible Experimenting G.O.D.-field process. This leads to the conclusion that all evolution, from the micro to the macro, must involve an expression of some sort of intelligent evolution.

Why do I say *"must involve"*? I choose these words using the identical reasoning that leads us to conclude that the spontaneous falling of apples to the ground *must involve* some sort of invisible guiding field, the one we call gravity.

Remember, no one has ever seen a gravitational field or ever measured one (except indirectly). We cannot hear, smell, taste, or touch gravity. What we do is logically infer the existence of a gravitational field from the behavior of apples falling to the ground, from planets revolving around suns, and light bending around the sun.

In the same way, no one has ever seen an Experimenting G.O.D.-field or measured one (except indirectly). We cannot hear, smell, taste, or touch an Experimenting G.O.D.-field. What we can do is logically infer its existence from everyday experiences such as the making of a sand painting or the folding of clothes (which we can prove requires an external, purposeful, intelligent process that can be seen by the naked eye) and, at another extreme, from the spiraling of galaxies and the creation of superclusters (which to the uninformed eye seem folded into spirals as if by magic). This leads to the conclusion that all evolution, from the micro to the macro, must involve an expression of some sort of intelligent evolution.

ASSUMPTIONS OF CONVENTIONAL SCIENCE AND OF THE SKEPTICS

The well-educated conventional science will be quick to retort that contemporary science operates on the premise that the evolution of order observed in the universe reflects a combination of (1) the operation of natural laws, plus (2) randomness, and therefore they see no need to hypothesize the existence of an invisible intelligence guiding the entire process. Such scientists will typically treat the sand painting and computer program examples described in Part Two as trivial (and will ignore them, since the examples clearly require the presence of intelligent Guiding-Organizing-Designing processes, human or otherwise). Instead they will focus their attention on the apparent "self" (species) evolution of plants and animals.

However, these scientists must assume that the discovered collection of so-called natural laws somehow came into being "by chance" (a presumption that is not based on any evidence—it is based on an implicit bias against intelligence rather than a careful analysis of intelligence) and that randomness, as in "inherently unpredictable," can somehow magically occur even in the absence of independence as is required for random sampling. I use the word "magically" only partly tongue-in-cheek, because such scientists simply assume that "inherent unpredictability" can occur; they do not say how it can occur.

If you are such a scientist (or a skeptic), we encourage you to remember that as explained earlier, the observation of "unpredictable" as in "I can't discover a pattern in these numbers" does not automatically justify your interpretation of the existence of an "unpredictable process" creating order out of "randomness." Humility about our assumptions, especially about randomness, is in order. The important lesson of the number pi is that it can remind all of us—skeptic, agnostic, and believer alike—that a seemingly "unpredictable" number can nonetheless be 100 percent replicable and reflect a complex, yet-to-be deciphered pattern of universal importance and application.

FROM INTEGRATION TO PREDICTION—CAN WE COMMUNICATE WITH AN INVISIBLE EXPERIMENTING G.O.D. PROCESS?

For an alternative theory to be valid it should first of all integrate diverse evidence, including evidence that seems anomalous to the conventional theory (for example, how the mind can literally organize seeming random electron flow, discussed in Chapter 15). The alternative theory should also make *new* predictions that can be confirmed or disproven (scientists prefer the term "disconfirmed") through future research.

When we integrate the concept of organizing fields from physics with the concepts of hierarchy and interconnections from systems science—what I have playfully termed "the All in the Small"—we come to the novel prediction that it is theoretically possible to receive information from a G.O.D. process, since the universal G.O.D. process is inside everything. In straightforward language, what this means is that both evidence and reasoning lead to the prediction that people can, to various degrees, receive information from "God." This includes you and me.

To the best of my knowledge, no credible researcher in the history of science has been brave enough to conduct a formal research program testing this prediction. The closest anyone has come, as far as I'm aware, was Emanuel Swedenborg, arguably Sweden's most successful and celebrated scientist. At age fifty-six he shifted his focus from the physical sciences to the spiritual sciences. The series of books he wrote in the 1600s were based upon the extensive personal exploratory experiments he conducted in his mind as he explored realms that could be experienced by human consciousness.

Swedenborg's personal exploratory experiments, shared through his numerous writings, have had a major influence on a select group of distinguished scientists and writers in the twentieth century, including William James and Helen Keller. However, no scientist has formally accepted the challenge of trying to test this prediction in the laboratory.

Swedenborg conducted theological experiments in his head for almost two decades. Einstein conducted thought experiments in his head beginning as early as his childhood. My personal exploratory experiments testing whether it is possible, in principle, to receive information from the universe that can be independently verified integrates Swedenborg's and Einstein's model of being open to exploring nature and the universe in one's mind.

The findings are not in and of themselves definitive, since by definition they are clearly exploratory. However, they are, as the Nobel Prize–winning physicist Wigner used to say, certainly "amusing" and "absolutely worth thinking about."

These preliminary findings provide proof-in-principle evidence— they demonstrate that it is possible to design controlled experiments in the laboratory testing the G.O.D. communication prediction. That is, if someone is brave enough to perform them. In part, the reason for writing this book is to provide the groundwork for bringing the Experimenting G.O.D. field process explanation into mainstream science so that an academic discipline that we might call "experimental theology" can be launched.

OTHER KINDS OF EVIDENCE–APPRECIATING THE EXTRAORDINARY SCOPE OF THE HUMAN MIND

There is a mathematical proof known as Gödel's Theorem that says, in essence, that certain mathematical assumptions can never be formally proven because of their complexity. Using the language of complexity theory, Gödel's Theorem concludes that to fully understand a complex system requires an even more complex system. A corollary of this idea from systems theory, as any youngster might say, is "It takes one to know one."

If human beings are composed of the very stuff of the Experimenting G.O.D. process itself—and therefore, if in principle we have its infinite potential to learn and evolve—then we should be able to find evidence in psychology and consciousness studies documenting that the human mind is at least as big in scope as the universe itself. (It is amusing that probably the most important mathematical proof of relevance to the Experimenting G.O.D. process is by Gödel, whose name happens to contain the word "God.")

In Chapter 12, I took you through a simple psychology experiment that convincingly demonstrates how even the mind of a young child can envision holding objects that vary in size from the infinitely small (a photon of light) to beyond the All (the entire universe) in the palm of their hand. The human mind has the power to learn how to invent technologies that can enable us to see the paths made by subatomic particles, on the one hand, and superclusters of galaxies, on the other hand. The human mind has the power to conceive of imaginary numbers and even infinities of numbers.

The history of science and technology reminds us that the potential of the human imagination goes beyond anything we can currently imagine. Thus far, no one has explained the origin of human consciousness in physical terms. Moreover, as described in Part Four, the experiments that purportedly support the thesis that the brain creates consciousness equally support the thesis that the brain is a receiver for consciousness. Experiments that distinguish between these two competing explanations—the brain as the creator of consciousness versus the brain as created by consciousness—are coming curiously from areas such as parapsychology and afterlife science that up to now have not been part of mainstream science.

It doesn't take a Nobel laureate physicist to reason that just as a raindrop is composed of billions of water molecules, and a lake is composed of billions of drops of water, and a galaxy is composed of billions of stars, and a universe is composed of billions of galaxies, so the universal Experimenting G.O.D. mind—if it exists as this book says it does—would be composed of billions of little experimenting Guiding-Organizing-Designing minds (little g.o.d.s) *using human brains.*

Part Five reviews evidence leading to the conclusion that just as invisible fields are universal, consciousness itself is universal. More-over, asking questions of the universe may be a universal process of intelligent trial-and-error questioning that exists at every level of the universe.

Evolution at all scales, from superstrings to superclusters of galaxies, and everything in between, including us, may not simply be "bottom up." It may well be "top down"—an example of "All in the Small" intentional and intelligent evolution of infinite scope and creative potential.

G.O.D. AND MATHEMATICS

Physicists have a passion, if not reverence, for mathematics. So, too, do deeply spiritual people, especially those who happen to have an affinity for math.

Einstein had a deep appreciation for the beauty and power of mathematics, as well as its limitations and his own limitations in dealing with it. He's quoted as saying "I don't believe in mathematics," as well as "Don't worry about your problems in mathematics; I assure you mine are greater."

On the other hand, there's some suggestion that Darwin may not have been a fan of mathematics. He has been quoted as saying, "A mathematician is a blind man in a dark room looking for a black cat that isn't there." As someone who has two cats, I appreciate both the utility and limitations of formulas and numbers—I appreciate both Einstein's and Darwin's complex opinions about math.

I love logic and mathematics, and I seem to have some degree of aptitude for both (despite receiving math awards in grade school, getting near perfect scores in new math when I was in high school, performing in the ninety-ninth percentile on the mathematics graduate record examination, and getting a hundred in a graduate course in complex analyses of variance, my strongest critics after reading this book would question whether I have a few mathematical screws loose—and I sometimes wonder this myself).

Some of my mathematically inclined colleagues, after reading early drafts of this work, have asked, "Where's the math?" They have argued, in various ways, that in order to believe that intelligence truly operates in physics, chemistry, and biology, they want to see the formulas. For them, words are ultimately fuzzy, whereas formulas are precise. When you transform an idea into an equation, the fuzzy speculation becomes a potentially testable premise.

(I will never forget when my agent, William Gladstone, who as an undergraduate at Yale was a math wizard, told me that what convinced him my theory of systemic memory had validity was not the three hundred pages of logic and data presented in *The Living Energy Universe,* but a single formula I included.)

So I offer a summary formula that speaks to the creation and evolution of feedback memory in systems—from superstrings to superclusters of galaxies, and everything in between, including us. It also speaks to the existence of intelligence and evolution in systems, since memory is a fundamental component of intelligence.

If you are not mathematically inclined, you can skip this formal equation and move on to a more verbal, poetic one; the abstract equation is included for those who are mathematically inclined. In the oversimplified formula described below:

- A and B stand for two parts of any system
- t stands for time (t+ stands for the time it takes information to travel from part A of the system to reach part B, or B to reach A)
- r stands for recurrence (repeating in a circular fashion, from A to B and back to A, over and over)
- "aba . . ." represents the evolving systemic (interactive) memory ($a_1 b_2 a_3$, etc.). The existence of nonlinear information circulating within a system allows for intelligence and creative emergence to operate within recurrent feedback processes.

Using these terms, the formula is:

$$(A_t + B_{t+})^r = (a_{t+} b_{t++} a_{t+++} \ldots)^r$$

To those who seek additional mathematical support (for example, how concepts of intention and purpose can be modeled as well), rather than offer it here in a book intended for the general reader, my colleagues and I will present the details in appropriate scientific publications, in the form of some novel equations and associated research.

A BIT OF POETRY

For those who prefer to skip the mathematics, I'd like to offer a related idea as a poetic formula, "a poetry of logical ideas"—a playful equation that expresses, in everyday language, the essence of the conclusion of this book. As Einstein said, "Pure mathematics is, in its way, the poetry of logical ideas." Physicists are often playful and poetic; the term "Big Bang" is a case in point.

In that vein, my playful expression of these concepts is:

$$\text{Evolution} = \text{motivation} \bullet \text{creativity}^{\text{Infinitely recurring}}$$

The take-home message of the intelligent evolution hypothesis is that motivation (another term for intention) and creativity (a component of infinite intelligence) together function as universal natural recurrent feedback processes that propel the origin and evolution of the order of everything—be it physical, chemical, biological, psychological, social, ecological, astrophysical, or whatever.

I'm not proposing that universal intention (motivation) and infinite intelligence (creativity) are human-like. I am not anthropomorphizing. What I'm proposing is that human intention and intelligence are universal-like, in the sense that our minds reflect the implicit infinite mind and feedback-guiding potential of the cosmos. Intention and intelligence may be as ubiquitous and universal as feedback. As stated previously, my proposition does not represent bottom-up logic (as below, so above) but rather top-down (as above, so below—a "meta" analysis).

As we've seen, order does not occur by chance; ordering mechanisms are presumed to manifest the replicable patterns we observe in the universe. Moreover, the apparent underlying complex order that defines the precise pattern of laws we have discovered in the universe implies the existence of a "meta-ordering" or "macro-ordering" process—what we have termed G.O.D.—that reflects a degree of extraordinary creativity and sense of purpose.

For the intelligent evolution hypothesis to be true, it must be sufficiently inclusive to (1) handle both the currently explainable as well as seemingly inexplicable observations (often termed anomalies) in nature and the laboratory, and (2) make new predictions that can be confirmed in future research.

As Einstein reminds us, there is more to reality than described by formulas: "As far as the laws of mathematics refer to reality, they are not certain; and as far as they are certain, they do not refer to reality."

SEEING THE BIG PICTURE

In Stephen Sondheim's haunting musical *Sunday in the Park with George,* about the Pointillist painter George Seurat, one of the songs includes the telling line "She saw all of the pieces, but none of the whole."

The pieces are all in place—from physics and statistics to psychology and parapsychology—and they are ready to be integrated. Will you be able to see the whole picture? And will you accept it when you see it? Compelling evidence for intelligent evolution and an experimenting G.O.D. process is staring at us from all directions; it is waiting to be integrated and witnessed as a whole. It appears that we no longer need historical religious figures to come to know God. We just need our rational and educated twenty-first-century minds, aided by the sacred gift of our innate intuitive potential.

All, everything that I understand, I understand only
because I love. Everything is, everything exists, only
because I love. Everything is united by it alone. Love is
God, and to die means that I, a particle of love, shall
return to the general and eternal source.

TOLSTOY

Infinite Love:
The Ultimate Gift from G.O.D.?

A few years ago, a woman whose husband had been brutally murdered came to see me at my office. Her husband had been an internationally recognized professor of music at the University of Arizona.
He was returning from an organ concert at night when he stopped to
help a man. It was raining. The man killed the professor for money to
purchase drugs. The woman, Stardust Johnson, had come to me because
my colleagues and I were conducting systematic research on the question of survival of consciousness after death. She, like her late husband,
was a deeply religious person and attended church regularly. Though
she desperately wanted to believe in the possibility of survival of consciousness after death, her anger at God was preventing her.

She said, "It's bad enough that God could allow my husband to be
taken from me for such a horrible reason"—for money to buy drugs.
Stardust believed that God gives us a degree of free will—the opportunity or chance to make up our own minds and behave accordingly—
and we too often pay a price for this liberty when people abuse their
freedom. Her belief is consistent with the science we have presented in
this book.

However, her subsequent words shook me to my core and I will

213

never forget them. She said, "However, if God has not only allowed my husband's love to be taken from me, *but he has allowed his love to be taken from me forever, such a God is too cruel to be imagined.*" I experienced her emotional and spiritual pain as if it were a sword in my soul.

When she said these words, I wondered to myself, What kind of a God would allow the light from distant stars to continue in the vacuum of space forever, but not allow our photons—our energy and information, our love and thoughts—to continue forever as well?

I responded to Stardust by asking her a question: "What would you say if I told you that well-accepted scientific theory supports the idea that our energy and information—and hence our love and thoughts—continue forever like the light from distant stars, and that it is possible to document this scientifically?" Stardust responded by saying, "It would give me a reason to hope and to live."

Is G.O.D. as envisioned by evidence-based faith "too cruel to be imagined"? Or have we simply not yet been able to fully envision the kind of loving G.O.D. process that actually exists?

SKEPTICISM, LOVE, AND KNOWING SOMETHING FOR SURE

Can the well-trained skeptic who is, so to speak, addicted to agnostic thinking (not meaning "believing it is impossible to know God" in the narrow religious sense, but meaning "questioning or wondering" in the broadest philosophical sense), know anything for sure?

Even when I describe things that happen 100 percent of the time—such as when I witness objects dropping to the floor or sand paintings scrambling in a pot—I describe these things as happening "virtually" beyond any doubt. I always include a little bit of doubt— keeping an open and discerning mind—for both logical and philosophical reasons. However, there is one thing that I know with absolute certainty.

It is fundamental to who I am.

When people ask me, "What do you know for sure?" I answer,

"The only thing I know for sure is that I love. I love people, animals, plants, places, books, music, art, sports, technology, stars, discoveries, and even ideas. My experience of love is beyond any doubt. Everything else I witness or experience I would describe as a hypothesis!"

Erich Fromm wrote that "love is the only sane and satisfactory answer to the problem of human existence."

When I experience love—especially deep love—it fills me completely. There is no doubt that I am having this experience. However, there is something absolutely foundational about love for me that to the best of my knowledge I have never heard described before.

And it relates to the question "Who is Gary Schwartz?" And by extension, who are you?

WHO IS GARY SCHWARTZ—IS HE WHAT HE HAS MANIFESTED, OR IS HE HIS YET UNMANIFESTED POTENTIAL?

Is Gary Schwartz a male who grew up on Long Island, played professional guitar in high school and college, became a Harvard Ph.D. in psychophysiology, conducted internationally acclaimed health psychology research at Harvard and Yale, directed an NIH-funded Center for Frontier Medicine in Biofield Science at the University of Arizona, writes science-based medicine and spirituality books for the general public, and has unbridled passion for the desert, Native American Art, ancient history, mysteries, bulldogs, and recently cats? Is this Gary Schwartz?

Or is Gary Schwartz the person that Gary could be? Am I my potential, the person I can be? And are you the potential of who you can be?

The fact is, you or I could have grown up in Kansas, and developed a Midwestern twang to our speech, or in Alabama. We could have grown up in France or China, speaking the language of the country. Though I began my undergraduate training as an electrical engineer, was premed and considered at least three majors—philosophy, psychology, and chemistry—I could have chosen any number of subjects to major in.

From this objective vantage point, I do not define myself in terms of my thoughts, philosophy, or history. I define myself in terms of not only who I am at the moment, but what my potential is. I see you the same way too.

My personal potential is of course limited—for example, it does not include my becoming a professional basketball player (I am physically too small) or a visual artist (I am red-green color-blind)—but that is not the point here. The point is that I define myself by my potential. I do not define myself simply in terms of what I have learned or achieved.

At times I have imagined what it might be like if my thoughts, philosophy, or history were taken away. What I experience when I do this thought experiment is that my essence—my potential—is still me. In other words, I can imagine losing aspects of my thoughts, philosophy, or history, and still feel like me. However, when I imagine what it might be like if my capacity to love were taken away, I experience myself disappearing in the process. If my capacity to love is taken away, my experience of Gary as I know him ceases to exist.

I have discussed this phenomenon with many people. They typically report having a similar experience of losing their essence when they imagine what it would be like if they completely lost their capacity to love. The capacity to love, and what to love, may define you, too, in a deep and foundational way.

THE REMARKABLE LOVING BRAIN

Humans are clearly hardwired for love. In fact, it can be argued that we come into the world as little global neurological love machines. As infants we find almost everything interesting—most things are curiosities to be tasted and examined. Depending upon our parents and upbringing, we will learn to either expand or contract our capacity to love.

Love can be stifled, suppressed, even imprisoned in anger and fear. Love can also be nurtured, encouraged, and fostered at every level. I was blessed to grow up in a home where exploration, love, and questioning were nurtured. My parents did not have money, but they had curiosity and an irrepressible passion for life. I was the fortunate recipient of that love of life, and I continue to feel it to this day.

I tell my students: "I love scorpions, sharks, rattlesnakes, and mountain lions; I just can't live with them. They are dangerous for me, so I keep my distance. But that does not mean that I can't appreciate their attributes, and marvel at their very existence." Loving many things does not necessarily mean that one lacks discernment or priorities. Also, one can love the capacity of animals or people to defend themselves and their loved ones, and hate how this capacity is sometimes abused by humans in the name of love.

It is said that God loves all creatures, great and small, be they on this planet or other planets in the universe. It is said that God has love and compassion for saints and sinners alike. It is said that God forgives our mistakes, and gives us the opportunity to try again and evolve into a being of love and light. Influential religious role models, both ancient and contemporary—from Moses and Jesus through Mother Teresa and the Dalai Lama—share one outstanding attribute. They have an overriding compassion and overabundance of love. They can discriminate right from wrong, yet also forgive those who have not learned the discrimination or mastered the ability to put the lessons into action.

Of all the known creatures on the planet earth, not only does the human mind stand out in terms of its ability for creativity and design, the human heart stands out in terms of its ability for love and compassion.

Just as our minds have the potential to explore all things, our hearts have the potential to love all things. The parallel here is nothing short of breathtaking. The evidence is overwhelming that our ability to love matches our ability to think. There are "heart geniuses" as there are "mind geniuses" (termed emotional and mental intelligence, respectively). The "Genius Within" is both a feeling genius and thinking genius.

But just as the individual mind must be developed to enable its potential to create to be actualized, the individual heart must be developed to enable its potential to love to be actualized. We must develop both our minds and our hearts. It behooves us to entertain seriously the possibility that the "Guiding-Organizing-Designing" process (mind) is also a "Generous-Opportunity-Developing" process (heart).

What I am proposing is that built into the very fabric of the universe is the capacity for infinite love and compassion as well as infinite intelligence and creativity. "What's love got to do with it?" My answer is: "Everything."

My working hypothesis, based upon available evidence in contemporary physics, is that the kind of God that would create the capacity for both visible and invisible photons never to lose their individuality in the "emptiness of space" is a God whose mind and heart should never be underestimated. Quite the contrary, I suggest that we seriously entertain the hypothesis that our ability to love, smile, and laugh is as vast as the universe itself and is as vast as the mind that can conceive it. Moreover, our love appears to be as permanent and expansive as the light from distant stars. Does that make you smile?

The question becomes, how far can we humans go in developing our capacity for love and compassion? It could be as great as our potential for creativity and understanding. Wouldn't that be the ultimate gift? I say we already have it.

Become the change you desire.

GANDHI

Be the miracle.

FROM THE FILM BRUCE ALMIGHTY,
WRITTEN BY STEVE KOREN ET AL.

Acknowledgments

I predicted this would be a daunting book to write; however, I did not fathom how challenging it would be. Writing the first draft was a thrill, but the process of revising, reorganizing, extending, and polishing the book evoked a broad range of emotional experiences, from agony to ecstasy. Were it not for the many people and the G.O.D. process who inspire me—knowingly or unknowingly—to reach beyond my formal academic training (electrical engineering, psychology, physiology) in order to attempt to answer the profound questions always bouncing in my mind, this book would never have been written.

First and foremost, I thank my writing partner, William L. Simon, for his encouragement, his probing questions, and his devotion to crafting prose that is clear and captures the essence of ideas. Bill is an open-minded skeptic who unfailingly asks the hardest questions. Occasionally he includes some easy ones as well. Though the content of this book primarily comes from me, the polishing of my words to make them clearer than I could is Bill's brilliant work. He and I are like hydrogen and oxygen that make water; when we come together and join forces, "emergent properties" arise where, as the saying goes, "the whole is greater than the sum of its parts."

Dr. James Levin, a deeply spiritual physician with a strong evidence-based faith in the methods of science, was the first person to encourage me to "tell these stories, tell them now, and tell them from your heart as well as your head." He introduced me to the writings of Joel Goldsmith, such as *God, the Substance of All Form* and *Consciousness Is What I Am.* Jim also introduced me to William Gladstone, whose editorial and agent skills are a blessing. It was Gladstone who first introduced me to Bill Simon and his remarkable wife, Arynne. Gladstone brought us to Atria Books, a publishing house that appreciates books that integrate science and spirituality. The staff at Atria, especially Brenda Copeland and LaMarr J. Bruce, our editors, have helped make this book "evocative and magnanimous" (LaMarr's words). Their commitment to high standards complements our own.

Jim also introduced me to Dr. Deepak Chopra. Deepak wrote the foreword to the previous book that Bill and I wrote, *The Afterlife Experiments.* I recommend Deepak's book *How to Know God,* which deserves to be read by anyone who wants to consider the question of all questions—the scientific reality of an evolving G.O.D.-field process in the universe as a whole.

The Guiding-Organizing-Designing hypothesis was seeded when I was a young assistant professor of psychology and social relations at Harvard University in the early 1970s and I read Dr. Walter Cannon's book *The Wisdom of the Body* and Alfred North Whitehead's *Process and Reality.* The concept of the "wisdom of the system" as in "the cosmos" came to me at that time.

The concept began to bloom when I was a young professor of psychology and psychiatry at Yale University in the early 1980s. There I read a combination of books that spanned science and religion, especially Dr. James G. Miller's *Living Systems,* Dr. Jonas Salk's *The Anatomy of Reality,* Dr. Lewis Thomas's *The Lives of a Cell,* Heinz Pagels's *The Cosmic Code,* Ervin Laszlo's *Introduction to Systems Philosophy,* Stephen Pepper's *World Hypotheses,* Harold Saxton Burr's *Blueprint for Immortality,* R. Wayne Kraft's *A Reason for Hope,* and Huston Smith's *The Religions of Man* (the original title, which has evolved into *The World's Religions,* a more appropriate, nonsexist title, as Linda Olds gently reminds me). Together they led to a transformation of my consciousness that continues to evolve to this day. I had the privilege to spend some quality time with both Salk and Pagels—their openness and insights hopefully live on in this book.

Because of contemporary books on chaos, self-organization, and complexity theory, such as Heinz-Otto Peitgen and colleagues' *Fractals for the Classroom,* the premise for this book has become ever more plausible. Thanks to remarkable colleagues such as Dr. Donald Watson, the rationale for positing some sort of ordering capacity in the universe—Watson proposes the novel integrative scientific term "enformy"—takes on even greater significance.

Glen Olds, Ph.D., and Eva Olds, Ph.D., a husband and wife team who have served many roles that bridge science, spirituality, and humanity—from ambassador to the United Nations and chair of the board of the World Federalist Association to the first president of the Fetzer Foundation and the president of three colleges and universities—encouraged me to be brave and stand up for the potential reality of a Guiding-Organizing-Designing process that science envisions to be a foundation of a living energy universe. Like me, they are quick to emphasize that the premise of this book is a scientific one—and like any hypothesis, it may be confirmed or disconfirmed through future research.

I thank Rustum Roy, Ph.D., an emeritus science professor from Pennsylvania State University and a member of the National Academy of Engineering, and William Tiller, Ph.D., an emeritus science professor from Stanford University and a leader in contemporary physics and consciousness research, for not only having served as honorary nonanonymous cochairs in the late 1990s and in early 2000 for the Human Energy Systems Laboratory's anonymous F.D.A. Committee (it stood for "Friendly Devil's Advocates"), but for bridging science, spirituality, and healing in their professional and personal lives. Some of the faculty members of our secret F.D.A. Committee feared that serving overtly in the role of critiquing some of the controversial research we are conducting in spiritual medicine and spiritual systems science might place their scientific careers in danger, hence our decision to make the committee anonymous. Thanks to chairmen Roy and Tiller, and to the invisible members of the committee, the scientific integrity of this work has hopefully been maintained at a high level.

I also thank Lynn Nadel, Ph.D., professor and former head, Department of Psychology; Holly Smith, Ph.D., former dean of the College of the Social and Behavioral Sciences; Al Kaszniak, Ph.D., professor and current head, Department of Psychology; and Richard Powell, Ph.D., former vice president for research—all at the University of Arizona—for fostering bridges between science and spirituality, and encouraging this work. Their understanding and respect for academic freedom makes this work possible in a university setting.

Integrative thinkers such as Ken Wilber and his "big bloom" vision, as described in his book *A Brief History of Everything,* have played an important background role in the birthing of *The G.O.D. Experiments.* You have met a number of integrative thinkers and their courageous ideas in the course of reading this book. And you have met some inspirational young people like Sabrina Geoffrion, M.A., who reminded me that systems theory predicts that we can all be in direct feedback with the G.O.D. process. I especially thank Lonnie Nelson, Ph.D., who revealed to me how conscious intention can lead the flow of electrons in an electromagnetically shielded device to become more organized, and then "forced" me to address the existence of universal consciousness in this book. I also thank Dan Lewis, B.A., who is part Cherokee; he revealed to me how even minerals and crystals have dynamically organized fields that can be detected consciously by humans under experimental, double-blind conditions.

Over the years my work has been funded by many federal sources, including the National Institutes for Mental Health and the Advanced Research Projects Agency (Harvard); the National Science Foundation and the National Heart, Lung, and Blood Institute (Yale); and the National Center for Complementary and Alternative Medicine, the National Institutes of Health, and the Research Office of Veterans Affairs (University of Arizona). Our research has also been funded by many private sources, including the Spencer Foundation, the Canyon Ranch

Foundation, the Peter Hayes Fund, and the John Kaspari Fund. Though none of this funding was directed toward the focus of this book, it provided the interdisciplinary education and integrative training I needed to do this work.

The help of my colleagues in our Center for Frontier Medicine in Biofield Science, funded by the National Center for Complementary Alternative Medicine from the National Institutes of Health, has been invaluable. Key members of my team, who have directly or indirectly contributed to this work, include Iris Bell, M.D., Ph.D.; Willow Sibert, Ph.D. pending; Allan Hamilton, M.D.; Lewis Mehl-Madrona, M.D., Ph.D.; Audrey Brooks, Ph.D.; Beverly Rubik, Ph.D.; Katherine Creath, Ph.D.; Katherine Burleson, M.D.; Melinda Connor, Ph.D.; Julie Beischel, Ph.D.; Maureen Campensino, Ph.D.; Cheryl Rittenbaugh, Ph.D.; Mikel Aicken, Ph.D.; Tomoe Lombard; Clarissa Siefert; Kim Corley; Katie Reece; Gerry Nangel; Sheryl Attig; Div. M.; Summer Stanwick; Larry Stephenson; and Jean Andrews. I cannot thank you enough.

I have received invaluable scientific and spiritual guidance from three special scholars and friends—Ernie Schloss, Ph.D., Jeanne Renouf, Ph.D., Ed.D., and Suzanne Mendelssohn, Ph.D. Hopefully this book honors your minds and hearts.

I have received invaluable critical feedback from two special physical scientists and friends, Peter Hayes, Ph.D., and Edgar Mitchell, Sc.D. Though we continue to disagree about the potential centrality of intelligence in the evolution of everything in the universe (for example, whether evolution is a bottom-up or top-down process), our debate has resulted in clarity as well as direction for future research.

However, in the spirit of brevity, I shall simply say bless you—to the thousands of authors, and many thousands of undergraduate and graduate students, colleagues, and friends, whose words and visions are an integral part of this work. And of course, if the thesis of this book turns out to be true, our blessings and dedication go to the living energy source of it all, the Universal Intelligent G.O.D.-field process who has so many names, including "Sam." We can thank him/her for intelligent evolution.

This book is also dedicated to the late Susy Smith—you are with me always, in my mind and heart. Not only does Susy inspire me, she also makes me giggle. And to Linda Van Dyck, Ph.D., my "spiritual interrogator"—you more than anyone have kept me focused on living a G.O.D.-conscious life.

So many teachers, so many gifts. Thank you all.

Gary E. Schwartz

Suggested Readings

The following are a small but important subset of books that contributed to the conceptualization and manifestation of *The G.O.D. Experiments* and the hypothesis of intelligent evolution. I list these twenty-six books with brief commentary about how they contribute to the evidence and the reasoning in this book. The books are organized by category. They are available from Amazon.com and other Internet sources.

ATHEIST VIEWS IN THE CONTEXT OF SCIENCE

Richard Dawkins's book *The Blind Watchmaker: Why the Evidence from Evolution Reveals a Universe Without Design* is a very clever book. He takes the idea of chance for granted. Using computer programs that require carefully designed feedback loops, Dawkins attempts to make the argument that if these programs can generate apparent design—they do—and Dawkins (the programmer) is not conscious of what the final outputs will look like, then this is compelling evidence for what he terms "The Blind Watchmaker." The problem with his logic is (1) he had to carefully design the programs to perform these purposeful functions in the first place (the programs were not created by chance), and (2) the fact that he could not foresee what the output would look like does not necessarily imply that someone smarter than he could not do so (humility is not Dawkins's strong suit). I suspect that most people who read his book will simply assume that chance operates in the universe. Also, they probably will not realize the fact that the programs written supposedly to disprove the existence of design actually require careful design on the part of the programmer. Dawkins's book is well written and worth reading even as it is misguided.

Michael Shermer's book *How We Believe: The Search for God in the Age of Science* is also a very clever book. Shermer assumes that science cannot address the God question experimentally, and that the evidence in favor of some sort of God hypothesis is weak at best. His book focuses on the psychology of how people can come to believe things for which there is no evidence, and also on how people can reach conclusions even though the evidence disproves their beliefs. Shermer's book is also well written and worth reading even as it is incorrect concerning the G.O.D. question.

Mark Perakh's book *Unintelligent Design* is one of the most thoughtful but misguided books on this subject. The description on the book jacket suggests that the work attempts to equate "seek[ing] to prove the existence of God mathematically" with "claims of neocreationism" and the views of "those who would see evolutionary theory discarded." In support of his position, Perakh challenges one of the most respected thinkers on the subject, the mathematician and philosopher Professor William Dembski, author of *The Design Inference*. Quoting Dembski's statement that "design . . . constitutes a logical rather than causal category," Perakh retorts, "If design is disconnected from any causal history, it seems to mean that Dembski's concept is that of a *design without a designer*." But Perakh misses another possibility: that Dembski means the appearance of design logically implies the existence of a designer, yet that comprehending a causal designer requires additional information. Though Perakh's book raises valid questions, the thoughtful reader will, I believe, likely come away unconvinced.

SCIENTISTS WHO REASON THAT SOME SORT OF INTELLIGENT EVOLUTION IS REQUIRED

Gerald L. Schroeder's book *The Science of God: The Convergence of Scientific and Biblical Wisdom* is a challenging book. Schroeder is a physicist who is deeply interested in the potential connections between contemporary physics and historical writings from the Bible. His book provides a substantial number of important, and sometimes surprising, parallels. It is a classic, filled with substantive and insightful information.

Lee M. Spetner's book *Not by Chance: Shattering the Modern Theory of Evolution* is a thought-provoking book. Spetner is an Israeli physicist and biophysicist. The book has been praised by scientists and religious scholars as one of the "most serious challenges to the modern theory of evolution." Professor E. Simon, Department of Biology, Purdue University, claims it is "certainly the most rational attack on evolution that I have ever read." Spetner's analysis and critique of neo-Darwinian theory (NDT) is based upon the mathematics of information theory. Spetner writes, "The NDT not only stands in the way of a better understanding of the life sciences, it also tends to prevent us from appreciating that there may be spiritual values in the universe that stem from a source higher than man."

Michael Behe, William A. Dembski, and Stephen C. Meyer, in their book *Science and Evidence for Design in the Universe,* provide an extensive review of scientific evidence consistent with the design hypothesis. Dembski reviews evidence for design in science in general, Meyer reviews evidence for design in physics and biology, and Behe reviews evidence for design at the foundation of life. They are some of the leading proponents of the intelligent design hypothesis, and it is important to understand the detailed evidence they review.

William A. Dembski's recent book *No Free Lunch: Why Specified Complexity Cannot Be Purchased Without Intelligence* is absolute must reading. Dembski has the training to envision this grand question. He has a B.A. in psychology, an M.S. in statistics, and a Ph.D. in philosophy. He also has a Ph.D. in mathematics from the University of Chicago and a Masters of Divinity from Princeton Theological Seminary. This book is not light reading; however, it is essential reading. Dembski reveals the fatal flaws in Dawkins's *The Blind Watchmaker* and makes a compelling case for the intelligent design hypothesis.

Guillermo Gonzalez and Jay W. Richards's book *The Privileged*

Planet: How Our Place in the Cosmos Is Designed for Discovery may be the most evidential of books concerning the design hypothesis. The question is raised, "Is the Earth merely an insignificant speck in a vast and meaningless universe," or is this "cherished assumption of materialism dead wrong?" The thesis of this book, filled with evidence from geophysics to astrophysics, is that "our planet is exquisitely fit not only to support life, but also to give us the best view of the universe, as if the Earth were designed both for life and for scientific discovery." This is a book that is definitely worth reading and pondering.

HOW CONTEMPORARY PHYSICS (OFTEN UNKNOWINGLY) TAKES US TO G.O.D.

Lynne McTaggart's book *The Field: The Quest for the Secret of the Universe* is a visionary book. It presents the latest theory and research on electromagnetic fields, biophotons, and the Zero Point Field. She shows how this research leads to the conclusion that an energy field connects everything in the universe, and we ourselves are part of this vast, dynamic network of energy exchange. For a reader-friendly introduction to field dynamics and contemporary physics, this book is excellent. However, implications for understanding God are not emphasized. The word "God" is not in the index.

Sten F. Odenwald's book *Patterns in the Void: Why Nothing Is Important* is a profound introduction to organized networks of invisible fields in the vacuum. Odenwald shares aspects of his spiritual pain with the reader (for example, he does not see how the new physics speaks to the question of survival of consciousness after death—even though physics provides such a foundation). However, his book is not focused on spiritual or theological questions. The word "God" is not in the index. But if you read this book with an open and discerning mind, the obvious take-home message about implications of the interconnectedness of everything, including the "void," is unavoidable.

Lawrence W. Fagg's little book *Electromagnetism and the Sacred* is beautiful and inspiring. Fagg is an interdisciplinary scholar. He holds a Ph.D. in physics from John Hopkins University and a master's degree in religion from George Washington University. Fagg explains that electromagnetism ultimately underlies all of earthly nature from rocks and plants to humans and their brains. He explains that "electromagnetic radiation—light—has symbolized divine presence in the spiritual life of humankind for millennia, and yet following the increasing disenchanting of nature, we have lost that contact between the physical and the spiritual." Fagg argues that the "ubiquity of electromagnetic phenomena constitutes a powerful physical analogy for the ubiquity of God's indwelling presence." The word "God" is heavily indexed.

William A. Tiller, Walter E. Dibble Jr., and Michael J. Kohane's book *Conscious Acts of Creation: The Emergence of a New Physics* is difficult reading if you are not a physicist. The book is pioneering and visionary. It "marks a sharp dividing line between old ways of scientific thought and old experimental protocols, wherein human quality of consciousness, intention, emotion, mind, and spirit cannot significantly affect physical reality, and a new paradigm wherein they can robustly do so." The book integrates contributions from medicine, parapsychology, consciousness studies, and especially physics. Tiller et al. discuss fields, the vacuum, space, and what they call the "Biobodysuit" metaphor. The word "God" is not in the index. However, their book provides compelling support for the concept of universal consciousness.

ASSORTED MEANINGFUL BOOKS ABOUT CONSCIOUSNESS, ORDER, AND INTELLIGENCE

Dean Radin's book *The Conscious Universe: The Scientific Truth About Psychic Phenonena* is essential reading. Radin not only reviews, in a definitive fashion, the experimental research on paranormal phenomena; he provides a wonderful introduction to how careful and creative research is conducted. He also discusses the numerous reasons why superskeptics dismiss the evidence as if it doesn't exist. The politics of science and truth are examined in this seminal synthesis of the field.

The Conscious Universe provides experimental evidence that strongly supports the existence of universal consciousness.

Tyler Volk's book *Metapatterns: Across Space, Time and Mind* is a meaningful book because he presents compelling evidence for the existence of repeating patterns of organization that are found at every level of the universe, from subatomic particles to galaxies and beyond. Not surprising, there are a few references to God in the index. One can't read this book and not be startled by the micro to macro organization of everything in the universe.

Mario Livio's book *The Golden Ratio: The Story of Phi, the World's Most Astonishing Number* is an astonishing book. Throughout history, thinkers from mathematicians to theologians have pondered the mysterious relationship between numbers and the nature of reality. Livio tells the tale of a number at the heart of that mystery, phi, or 1.6180339887 .... The golden ratio shows a propensity to appear in the most remarkable places—from mollusk shells, sunflower florets, and the crystals of some materials to the shapes of galaxies containing billions of stars. From art and architecture to economics and astrophysics, the phi metapattern is discovered to be a beautiful and seemingly eternal order.

Frank T. Vertosick Jr.'s book *The Genius Within: Discovering the Intelligence of Every Living Thing* is surprisingly enlightening. Vertosick explains how our bodies contain many highly intelligent systems just as remarkable as our brains. We come to respect "the brilliance of viruses and the canniness of germs." I recommend this book because it helps one understand the essence of intelligence. When you can see how the immune system works with "a savvy and devious cleverness," you are on the path to seeing intelligence in all dynamical systems, not just biological systems. Though the word "God" is not in the index, the title of this book implies his focus.

Gary E. Schwartz's book (with William L. Simon) *The Afterlife Experiments: Breakthrough Scientific Evidence for Life After Death* reviews contemporary research conducted at the Human Energy Systems Laboratory at the University of Arizona that speaks to the question "What comes first, brain or mind?" The research provides compelling evidence not only for the existence of a larger spiritual reality, and that human consciousness continues "like the light from distant stars," but also that consciousness is primary in the universe, and that the physical world, including the human brain, is a manifestation of this universal process.

BOOKS ABOUT GOD, CONSCIOUSNESS, EVOLUTION, AND RELIGION

Huston Smith's book *Why Religion Matters: The Fate of the Human Spirit in an Age of Disbelief* is one of his most important works. An eloquent and respected authority on religion, Smith offers a timely manifesto on the urgent need to restore the role of religion as the primary humanizing force for individuals and society. Weaving together insights from comparative religions, theology, philosophy, science, and history, along with examples drawn from current events and his own extraordinary personal experiences, Smith gives both a convincing historical and social critique and a profound expression of hope for the spiritual condition of humanity. He takes both historical religion and science to task. In the final part of his book, Smith imagines a time when "human beings move beyond the present materialistic and relativistic understanding of existence and recognize that consciousness, not matter, is the ultimate foundation of the universe."

Jeff Levin's book *God, Faith, and Health: Exploring the Spirituality-Healing Connection* provides an empirical reason why believing in God and sharing this experience with others matters. Levin explores the latest compelling evidence of the connection between health and an array of spiritual beliefs and practices, including prayer, attending religious services, meditation, faith in God, and others. With examples from spiritual traditions as diverse as Christianity, Judaism, and yoga, he looks with an open mind and perceptive eye at the many ways that religious involvement and belief can prevent illness and promote health and well-being.

James Redfield, Michael Murphy, and Sylvia Timbers's book *God and the Evolving Universe: The Next Step in Personal Evolution* is both historical and visionary. The authors contend that thousands of years of human striving have delivered us to this very moment, in which each act of self-development is creating a new stage in planetary evolution—and the emergence of a human species possessed of vastly expanded spiritual experience. One important discussion is their chapter "The Experience of Integration and Synchronistic Flow."

Deepak Chopra's book *How to Know God: The Soul's Journey into the Mystery of Mysteries* is a tour de force. According to Chopra, the brain is "hardwired to know God." He reviews how the human nervous system has seven biological responses that correspond to divine experience. These are shaped not by any one religion (they are shared by all faiths), but by the brain's need to take "an infinite, chaotic universe and find meaning in it." Chopra says that "God is our highest instinct to know ourselves." His book integrates contemporary physics, physiology, and parapsychology with seven levels of the experience of God.

George Trobridge's biography *Swedenborg: Life and Teachings* is arguably the most authoritative history of Emanuel Swedenborg, an eighteenth-century scientist considered to be an "intellectual titan." Swedenborg studied philosophy, science, mathematics, law, Latin, Greek, Hebrew, cosmology, anatomy, physiology, politics, economics, metallurgy, mineralogy, geology, mining, engineering, and chemistry (he did not study psychology or sociology; these disciplines did not exist in his time). His studies ultimately led him to a search for the physiological seat of the soul. Following his investigations in the natural sciences, which gained him prominence throughout academic Europe, he dedicated his last twenty-seven years to theology and the study of the Bible. In this final career phase, he believed himself to be "under God's call and illumination." A true empirical scientist and experimental theologian, his life and teachings are worth knowing. His writings influenced the lives and works of many noted personalities, including Kant, Emerson, Henry and William James, Dostoyevsky, Strindberg, Jung, and the celebrated Helen Keller.

Michael Reagan's *The Hand of God: Thoughts and Images Reflecting the Spirit of the Universe* is a beautiful book. This book combines inspiration for the mind and heart by juxtaposing majestic photographs of the cosmos with illuminating words of scientists, poets, and theologians. I was introduced to the integrative spirituality of Wernher von Braun in this book. Here's how Chet Raymo, author of *Skeptics and True Believers,* is quoted in the book: "Our explorations have produced a vast archive of remarkable astronomical images. . . . The riches are too many for choices, the revelations beautiful and dreadful. Who can look at these images and not be transformed? The heavens declare God's glory." One more quotation in the book, from Gerald L. Schroeder, author of *The Science of God:* "Knowing the plumbing of the universe, intricate and awe-inspiring though that plumbing might be, is a far cry from discovering its purpose."

Gregg Braden's *The God Code: The Secret of Our Past, the Promise of our Future* is a bold, controversial book. Braden takes us on a journey of discovery, leading us to his profound hypothesis of a universal code that is expressed, in parallel, in the original Hebrew language related to God and in our DNA code. According to Braden's logic, the basic elements of DNA—hydrogen, nitrogen, oxygen, and carbon—directly translate into specific letters of the Hebrew alphabet (*YHVA*), which then translate into the original name of God. Braden's hope is that knowing that God's signature is carried within each cell of the estimated 6 billion humans on earth will give humankind the evidence we need to overcome our differences and renew our faith. He writes, "Beyond Christian, Jew, Muslim, Hindu, Buddhist, Shinto, Native, Aboriginal, white, black, red, or yellow; man, woman, or child, the message reminds us that we are human. As humans, we share the same ancestors and exist as the children of the same Creator. In the moments that we doubt this one immutable truth, we need look no further than the cells of our body to be reminded. This is the power of the message within our cells." Is this simply a statistical accident, or further evidence of a Guiding, Organizing Designer?

Neale Donald Walsch's *Tomorrow's God: Our Greatest Spiritual Challenge* is also a tour de force. It integrates ancient and contemporary sacred wisdom in a way that is folksy, down-to-earth, and easy to digest, and at the same time is exquisitely visionary. He writes in the introduction that the purpose of writing this book was "not that it contains wisdom that we have not been given, *but that it repeats wisdom to which we have not been listening*" (italics are Walsch's). He goes on to say, "And the problem is, if we do not listen to this wisdom now, we may not have many more chances to have it repeated. We are at the edge, you see. We have gone as far as we can go in the direction we have been taking. We need now to change course if we wish to preserve life as we know it on this planet." Walsch proposes that each of us can contribute to changing "the course of human history." Is this an exaggeration, or an achievable goal? Walsch shows how evolving our vision of God is key to our evolution as a species.

Ted Peters and Martinez Hewlett's book *Evolution from Creation to New Creation: Conflict, Conversation, and Convergence* provides an important overview and integration of theology and science. Their synthesis combines the knowledge of a theologian (Dr. Peters) with a biologist's (Dr. Hewlett). The conclusion they have reached is summarized in the term "theistic evolution"—which overlaps strongly with the more generic term "intelligent evolution" as used in *The G.O.D. Experiments*. In a 2005 opinion article, they wrote that their concept of a God who "created our world in the beginning and continues to provide creative and redemptive care [is] compatible with scientific methods that are blind to divine purpose within natural processes." Further, they wrote that "what natural scientists study we call 'secondary causes,' whereas God is the 'primary' cause. Science can contribute immensely to understanding the created world replete with secondary causes, and indirectly contributes to our understanding of the creator, the primary cause." At the present time, there is sufficient empirical (evidential) and conceptual (theoretical) reason to hypothesize the existence of "theistic evolution"—which is why Peters and Hewlett wrote their book and teach contemporary evolutionary theory from this integrative perspective. People committed to honesty and integrity in education will hopefully encourage science teachers to consider teaching the hypothesis of theistic evolution—or more broadly, intelligent evolution—with their students.

Though I agree in principle with Peters and Hewlett's cautious recommendations about theistic evolution and education, I would remind the reader of the essential distinction between observation (evidence) and interpretation (theory or models). Science is concerned both with data (including seemingly anomalous data) and alternative explanations that can account for the data. At the present time, there is sufficient empirical (evidential) and conceptual (theoretical) reason to hypothesize the existence of "theistic evolution"—which is why Peters and Hewlett wrote their book and teach contemporary evolutionary theory. People committed to honesty and integrity in education will hopefully encourage science teachers to consider sharing the hypothesis of theistic evolution (or more broadly, intelligent evolution) with their students.

APPENDIX A

A Review and Study Guide for Teachers and the Rest of Us

The G.O.D. Experiments presents a body of evidence from physics, mathematics, psychology, and parapsychology that provides compelling reasons to posit the existence of intelligence everywhere present in the universe and daily life. The book presents substantial evidence combined with systematic reasoning for the existence of an intelligent, trial-and-error, universal G.O.D. process—a Guiding-Organizing-Designing process. The approach combines elements of evolution theory with aspects of intelligent design theory; for ease of communication, the approach is called intelligent evolution.

The evidence not only crosses the "beyond reasonable doubt" threshold, but some of it virtually crosses the "beyond any doubt" threshold as well.

The *Introduction* poses the question "Can science per se lead us to discover some sort of a G.O.D. process?"

Parts One–Three present the evidence in detail, spanning from physics to parapsychology.

Part Four addresses the theoretical debate about chance versus design in the universe, and comes to the conclusion that chance per se cannot explain the origin and evolution of order in the universe. Logic and evidence point to some sort of intelligent evolution process.

Part Five plus the *Epilogue* discuss some of the numerous implications of adopting evidence-based faith concerning the existence of a G.O.D. process.

Hopefully this review and study guide will provide teachers a means of helping their students see both the forest and the trees—the big picture plus the important details—as they evaluate the evidence and come to their own conclusions. And hopefully it will, as well, provide interested readers a useful summary overview as a refresher of the core ideas and arguments.

The reader will note that this is neither a religious nor a political book—it is a science book. It describes a process of discovery leading to a challenging conclusion that has the potential to evolve both science and religion. The synthesis is termed intelligent evolution.

PROLOGUE. CAN SCIENCE TAKE US TO GOD?

Imagine that there had been no Abraham, no Moses, no Jesus, and no Muhammad, no Torah, no Bible, and no Koran, no shamans, no medicine men, no prophets, and no mystics. There was just contemporary, twenty-first-century science.

Would scientists, on their own, come to the conclusion that some kind of invisible G.O.D.—a Guiding-Organizing-Designing process—existed in the universe as well as in our daily lives?

There is no bigger or more important idea than the possible existence of a universal, caring, creative, intelligent Guiding-Organizing-Designing process that helps form and guide the evolving macro system we call the universe.

Can the G.O.D. process question be addressed by contemporary science?

The Prologue introduces this monumental question. It briefly reviews (1) how the seed of the idea came to the author when he was a young assistant professor of personality psychology at Harvard University in the early 1970s, (2) how it blossomed when he was a professor of psychology and psychiatry at Yale University in the 1980s, and (3) how it evolved into a veritable orchard when he was a professor of psychology, medicine, surgery, neurology, and psychiatry at the University of Arizona in the 1990s.

The concept of evidence-based faith—which at first may sound like a contradiction in terms—is introduced in the Prologue.

It is proposed not only that science is solving what Deepak Chopra in his book *How to Know God* calls the "mystery of mysteries," but more important, that science is serving the mystery of mysteries in a manner that portends our capacity for personal and global healing—including the healing of our business, political, and religious institutions.

The Prologue explains that this is not a religious book examining ideas about "God"—it is a science book about a universal G.O.D. process that addresses the common ideas of higher purpose, higher intelligence, and higher power in the universe and everyday life.

PART ONE—THE SCIENCE OF PROPHECY

CHAPTER 1. FORESEEING GOD IN THE LABORATORY

Chapter 1 begins with an overview of an extraordinary experiment testing Christopher Robinson, England's Dream Detective. The ten-day experiment involved Robinson dreaming about specific locations he would be taken to in the greater Tucson area on the following day. The results from one particular day are described in detail.

The dream information included the phrase "the spirit of God." At the end of the day, the author and Robinson found themselves eating under an umbrella displaying the advertisement for Hebrew National hot dogs; the line on the umbrella read: "We answer to a Higher Authority."

How could Robinson know not only specific details about where the experimental location would be, but the surprising connection with God?

Could this be explained by chance per se? Or was there some sort of an invisible, intelligent Guiding-Organizing-Designing process operating in our lives (and by extension, yours)?

CHAPTER 2. DISCOVERING INTELLIGENT DESIGN IN OUR LIVES

Chapter 2 presents unanticipated evidence for the existence of intelligently orchestrated design in daily life. As the Ten-Days-in-Arizona experiment unfolded, the author began to realize that certain pieces of information that Robinson was getting from his dreams in May, June, and July were predicting the existence of activities, as well as the coordination of these activities, that happened later in August.

Day 10 was especially prophetic—it happened to fall on a day that the author had to give a lecture in the early afternoon at the University of Arizona's College of Medicine. The dream Robinson had in June said that "Day 10 will end early, and you will spend the day at the pool." Even though (1) the actual date for the experiment had not been set in June, (2) the author had not selected possible sites to visit, and (3) the sites ultimately selected had not yet been randomized, Robinson's dream accurately foresaw the author's schedule as being integrated with his.

The building visited on Day 10—a Native American museum—turned out not only to be a few blocks from Robinson's hotel, but to be closed that day. As a result, the author had more than enough time to get to the medical school by noon, and Robinson did spend most of the day by the pool.

What kind of superintelligence can foresee orchestrated events that have yet to be selected and randomized by humans? Such intelligence is by all definitions "supernormal" if we use human intelligence as the standard of normal.

PART TWO—SIMPLE G.O.D. EXPERIMENTS

CHAPTER 3. K.I.S.S.: KEEP IT SIMPLE SCIENCE

This chapter demonstrates a simple experiment that families can perform in their homes to show how the predictions of chance per se don't work when they are actually tested. The experiment involves sand paintings and what I call the "G.O.D. container" experiment.

Take a container, place white sand in the bottom, and then purposely draw something using black sand. You can draw a cross, a Star of David, a heart, a human eye, the word "G.O.D."—it does not matter. Then, cover the container—preferably with a clear cover so you can see what happens—and shake the container. What happens? Does the cross become a star? Does the heart become an eye? No, what happens is universal and occurs 100 percent of the time. What the sand does is mix. The black pieces spread out and uniformly mix with the white sand. The more you mix the sand, the more the black and white pieces distribute themselves more or less equally.

The take-home message is significant. Sand paintings do not happen by chance. No amount of time will lead the sand to paint itself, by chance alone.

This is evidence, not theory.

Order in the simple physical world does not occur by chance. It takes some sort of a Guiding-Organizing-Designing process to paint a sand painting (or construct a building); it takes some sort of intelligent ordering process to create and maintain ever more complex orders that we witness in an evolving cosmos.

CHAPTER 4. G.O.D. IN THE COMPUTER

This chapter examines whether randomness can occur by chance. What statistics tells us is that the more data we collect—for example, the more times we flip a coin—in the absence of a Guiding-Organizing-Designing process, the greater is the likelihood that a bell-shaped, "normal" distribution will occur.

In the case of coin flips, this means that the greater the number of times a coin is flipped—and no ordering process (or cheating) is involved—the greater the probability that the average will approach 50 percent heads and 50 percent tails. Similarly, the more time we give monkeys the opportunity to peck at the keys on a piano, the greater the probability that they will hit the notes equally, and therefore will never compose a Bach symphony (or even orchestrate "Twinkle, twinkle, little star").

The deep implication of this is that the normal curve—its shape and structure—will always occur in the absence of a nonrandom process that could shift the numbers away from chance.

Chance occurs only when each and every event is independent of every other event. This is called random sampling. When you set up these precise conditions, the more data you collect, the more the distribution produces exactly the same normal shape.

The bell-shaped curve never becomes triangle shaped, or square shaped, or inverted. And extreme numbers—such as one hundred heads in a hundred flips—virtually never occurs.

This is not theory; it is evidence—it is fact. You can program your personal computer to generate random numbers and plot the distributions. The normal curve shows up every single time. It is as reliable as sand mixing in a container.

In the presence of complete independence, chance does not create improbable novel orders. It creates normal curves.

The explanation and surprising purpose for this fundamental observation is provided in Part Four.

Clearly, chance could neither explain nor create the kinds of life patterns observed with Robinson in Part One, let alone the existence of supermacro organized patterns of billions of spiral galaxies in the universe.

PART THREE—CAN YOU COMMUNICATE WITH G.O.D.?

CHAPTER 5. TALKING TO AN INTELLIGENT BLACK BOX

Part Two indicates that chance cannot explain the existence of order in the Dream Detective experiments described in Part One. Moreover, Part Two indicates that when chance is allowed to operate, one always gets a mixture—be it of sand or of numbers.

If chance does not, by itself, produce the origin and evolution of complex orders (Part Two), and if exquisitely orchestrated orders exist subliminally in our lives (Part One), then it should be possible, in principle, to receive information from this Guiding-Organizing-Designing process to help navigate our lives.

Chapter 5 provides the theoretical rationale from systems science that explains how "the All is in the Small"—how a G.O.D. process is theoretically present in all things, from the micro to the macro. The fictional story of an intelligent black box is shared, with a surprise ending (you'll have to read the chapter for the ending) that indicates how we must be willing to ask the right questions, with the right intentions (including genuineness and humility), if our goal is to receive the right answers.

Chapter 6. I Asked the Universe a Question

In Chapter 6 the author performed an exploratory experiment and asked the universe a question. He was a professor at Yale at the time, and was beginning to accept the possibility that everyone, in principle, had the possibility of receiving information from a higher or universal creative intelligence.

The author hated the word "God" (his childhood image was of a white bushy-bearded man with a cane who went around spreading plagues on people). He decided one night to ask the universe if it would give him a new name for "God." The surprising name he received (you'll need to read the chapter to learn the name) initially sounded ridiculous but turned out to have a Hebrew root that means "the name of God."

The author considered eleven possible explanations for the word he received, beginning with the possibility that he learned the answer when he went to temple as a child. However, the only explanation that fit the totality of the history and experience was the eleventh explanation: "Be careful what you ask for."

The author was sufficiently frightened by the potential implications of his initial personal experiment that he did not ask another question of the universe for close to ten years.

Chapter 7. I Asked the Universe More Questions

Lest the reader conclude that the name the author received in response to his question was simply a lucky guess (one of the eleven possible explanations considered in Chapter 6), the author includes one more example of an additional personal experiment where he asked the universe a question.

This question was triggered by a woman who boldly asked him a question after he gave an invited address at a science-of-consciousness meeting in New Mexico. The woman asked him, "Do you believe in the theory and data you presented?"

The author attempted to explain to her, and the audience, that scientists do not "believe"; what they do is consider alternative hypotheses and evaluate the probability that a given hypothesis is true.

The woman was not satisfied with his answer. She repeated her question more vehemently; he proposed they talk about it later.

That night, in his hotel, the author decided to ask the universe the following question: "How can I know—and prove—that some sort of a Guiding-Organizing-Designing field exists in the universe?" (He sometimes refers to this as the G.O.D.-field process.)

The answer he heard completely surprised him. It was "Remember the diamond." This was followed by three images of diamonds, from the origin of diamonds in the earth, to a skilled cutter shaping the rough diamond, to the use of the diamond in the COBE space telescope to detect the nonrandom background radiation that occurred shortly after the Big Bang.

The following morning, an extraordinary coincidence occurred that was as remarkable as any coincidence observed in the Robinson experiments (you will have to read Chapter 7 to learn the coincidence). The event insured not only that the author would never forget the experience, but that he would ultimately confess the experience as personal evidence for some sort of a Guiding-Organizing-Designing process in the universe.

CHAPTER 8. INTERESTING AND AMUSING THEORIES

What is the process by which scientists search for evidence and draw conclusions from data, especially if the conclusions are bold and far-reaching?

Using the phrase from emergency room medicine "When you

hear hoofbeats, don't think zebras," the chapter reviews how responsible medicine and science adopt the framework of considering the most probable explanations first (for example, in the United States, horses), and consider the less probable explanations (for example, zebras) only after the most probable explanations have been ruled out.

If we use the criteria "beyond reasonable doubt" and even "virtually beyond any doubt," the totality of the evidence from physics, mathematics, psychology, and parapsychology leads to a simple and parsimonious conclusion—called Ockham's Razor.

The horse/chance explanation does not account for the totality of the data; a zebra/universal intelligence explanation is required. In a word, the thesis of this book is "When you hear hoofbeats, think a G.O.D. process."

Chapter 8 also discusses the difference between "interesting" and "amusing" theories. The Nobel Prize–winning physicist Eugene Wigner, of Princeton University, taught his students that "interesting theories, though often true, are often not worth thinking about, whereas amusing theories, though often not true, *are absolutely worth thinking about*." The book illustrates how the proposed Experimental G.O.D. process theory, existing as Universal Consciousness, fits the criteria of being an "amusing" theory worthy of our serious contemplation. It is important to remember that sometimes amusing theories turn out to be true.

The chapter ends with examining how people vary in their beliefs regarding some sort of a G.O.D. process—from strong believers, to agnostics, to extreme disbelievers.

This book is not about culture-based beliefs derived from religion; it is about evidence-based beliefs derived from science. The evidence is provided in such a way that the reader not only can make up his or her own mind, but in the process, can develop insight into how he or she comes to reach these conclusions—or why she or he resists them.

PART FOUR—THE GREAT G.O.D. DEBATE

CHAPTER 9. CHANCE VERSUS INTELLIGENT DESIGN—WHICH IS IT?

Chapter 9 discusses the "chance universe" versus the "intelligently designed universe" debate in some depth. It reviews the debate from both a philosophical and scientific point of view.

In the process of reviewing the great debate, the fundamental but often forgotten distinction is drawn between observations (evidence) and interpretations (explanations), using the Copernican Revolution as an illustration.

Throughout recorded history—up to the present day—humans on the earth have witnessed the sun seemingly rise in the east and set in the west. In Tucson, which typically receives more than 330 days of clear skies a year, it is possible to see a bright yellow round-shaped object appear seemingly out of nowhere in the east. It then seems to move across the sky during the day and ultimately disappears at the end of the day.

This completely replicable observation led people for thousands of years—including distinguished scientists and religious leaders—to the interpretation/explanation/conclusion that the sun revolves around the earth.

It is now well known that this conclusion is a mistaken interpretation of the observations. There is an alternative explanation for the relationship between the sun and the earth that creates the same observations of the sun seemingly revolving around the earth.

Copernicus predicted, and Galileo confirmed, that the alternative and highly controversial interpretation—that the earth revolves on its axis and circles the sun—was the more accurate and probable explanation. Not only did the Copernican "heliocentric" explanation provide a simpler description of the observations than the commonsense "geocentric" explanation; it predicted and explained seemingly anomalous observations in the movements of stars that could not be handled by the geocentric theory.

The take-home message from this paradigm-changing moment in the history of science is that just because the sun looks like it is revolving around the earth—an interpretation of the observations—does not make it the only interpretation or the correct one.

This book proposes that the same lesson applies to coin flips and our interpretation of chance. There is no question that under certain conditions—complete independence of events—we will observe approximately 50 percent heads and 50 percent tails. However, this does not mean that the chance or random interpretation/explanation/conclusion is either the only conclusion or the correct one. The author suggests that there is a new revolution in the interpretation of apparent randomness and chance in the universe.

Chapter 10. Can G.O.D. Play Dice with the Universe?

The new, revolutionary explanation for the observation of coin flips is actually quite simple; in fact, it is just as simple, in principle, as is Copernicus's revolutionary explanation for the observation of the sun's relationship to the earth.

Recall that to create the conditions necessary to observe chance, each event must be independent of every other event. Moreover, each event must be the same as the previous event—it cannot change, grow, or learn over time. Under these conditions, a normal distribution will be observed every single time, especially if you include more coin flips or data.

The question then arises, does nature exist this way—does it fit these required conditions? The answer is plain and simple: no. What physics tells us is that events are not independent; they are interdependent to various degrees because everything is interconnected by gravitational, electromagnetic, and quantum fields.

The first person to state this expressly was Sir Isaac Newton with his theory of gravity. Newton proposed that every physical object, regardless of its size, has mass that pulls in all directions. This gravitational force extends into space, in all directions—and according to contemporary physics, it extends infinitely!

This means that not only does the earth pull on you and the moon; you pull on the earth and the moon. The moon cannot move without affecting the earth and you; conversely, you can't move without affecting the earth and the moon. Of course, your mass is much smaller than that of the earth or the moon, so your effect is less intense. But the effect is still there.

This is why when you throw a coin in the air and you are standing on the earth, it falls. Moreover, it does not fall randomly, because it is influenced, to various degrees, by all objects in its vicinity, including you and other people around you—even though it averages out to approximately 50 percent.

Newton was an extremely religious man; he saw the universal, invisible, unconditional, nonprejudicial, attractive force that held the universe together to be an expression of "God's love" in the universe.

If the conditions of independence do not ultimately exist in nature, then the distributions observed in nature should never actually be normal—they will be skewed or shaped to various degrees by the organizing forces interconnecting the objects.

In Pagels's revealing chapter on randomness in his book *The Cosmic Code,* he lamented that physicists, when they looked closely, discovered that the patterns of numbers observed in nature *never* expressed true randomness as produced by random sampling. Like most physicists, he found this quite puzzling. I underscore "never."

However, if interdependence is the rule, not the exception, in the universe, then the conditions necessary to produce a normal distribution are not present in the universe. It logically follows that we cannot, with integrity, use chance or randomness as an explanation for the origin and evolution of order in the universe if the conditions necessary to produce randomness are not present in the first place.

If chance cannot be an explanation for order, then the existence of some sort of Guiding-Organizing-Designing process must be hypothesized.

If intelligence is the rule, not the exception, in the universe, then what intelligent purpose would apparent randomness (events that appear to be unpredictable) serve if it was not really random?

Why would a universal intelligent designing process make it possible for apparent randomness to exist in the first place? Could this be an expression of intelligent evolution?

A simple K.I.S.S. explanation suggests itself. Since randomness—

as produced by random sampling—does not occur by chance, *what apparent randomness does is give, within limits, the opportunity or chance for novel orders to occur.*

In other words, relative randomness is a tool for allowing creatively designed orders to occur. Relative randomness allows for intelligent, trial-and-error learning to occur in the evolving universe.

Therefore, the new revolution in thinking is the hypothesis that chance provides the opportunity for creatively designed orders to be discovered and expressed.

In a sense, God does play dice with the universe; however, the purpose is different than we thought historically and it is ultimately consistent with Einstein's spiritual vision.

INTERLUDE. THE "DIVINE PROPORTION"

One of the most remarkable metapatterns that replicate throughout nature and the universe is defined by the numerical ratio termed phi. Phi is defined by mathematicians to be an irrational number—it is like the pi ratio in that no matter how far its digits have been extended, we have never discovered a repetition that defines its order. Because we are currently not able to discover its order, we typically assume (mistakenly) that it must be random or disordered.

What's truly astonishing about the phi ratio is that it pops up in diverse shapes and dynamics spanning art and architecture, botany and biology, physics and mathematics, and even economics and astrophysics. The golden ratio is found to occur in the dynamics of the stock market as well as the shape of billions of stars in a spiraling galaxy.

Does phi, sometimes called the golden ratio, reflect a "Divine Proportion"? Is G.O.D., among other things, a universal and infinite mathematician? These age-old questions take on new meaning in light of the evidence and reasoning presented in this book.

CHAPTER 11. WHY SCIENCE SHAVES WITH OCKHAM'S RAZOR

Chance, by itself, cannot explain the origin and evolution of order in the universe. So does this necessarily require that we posit some kind of intelligent designing process to explain everything that exists in the universe?

The key word here is "everything" and the commonsense answer is "No."

Even if we accept the fact that sand paintings never paint themselves, or that watches never assemble themselves, we recognize that many phenomena in nature *seem* to organize themselves quite well. I underscore "seem."

Everyday experience reminds us that clouds appear to spontaneously emerge, raindrops appear to liquefy and fall naturally, and tornadoes appear to swirl and create destruction, all seemingly by themselves. Oak seeds appear to grow into oak trees, and fertilized human eggs appear to grow into newborn babies, again all seemingly by themselves.

Phenomena that seemingly invent and evolve themselves are scientifically termed "self-organizing systems." Nature is replete with systems that show apparent self-organizing properties.

This chapter addresses this question from the perspective of Ockham's Razor. It is a fact that some systems appear as if they self-organize; it is also a fact that the earth appears to be flat. The key term in both cases is "appear." One approach is to posit that there are two classes of phenomena in nature, one class or set of evidence that requires some sort of an external Guiding-Organizing-Designing process (e.g., sand paintings), another class or set of evidence that seems not to (e.g., clouds). A second approach is to include and integrate all of the evidence—including the existence of invisible fields that interconnect everything that exists in the universe—and posit that a single explanation, the G.O.D. process explanation, accounts for both classes of phenomena. This is the approach of Ockham's Razor.

The G.O.D. process explanation also makes new predictions that can be confirmed or disconfirmed in future research. The organizing sun prediction is explained as a logical extension of using Ockham's Razor and applying it to the G.O.D. process explanation.

PART FIVE—IMPLICATIONS OF G.O.D. FOR EVERYTHING

CHAPTER 12. WHO ARE WE AND WHY ARE WE HERE?

Chapter 12 returns to the power and potential of the human mind, and illustrates how the human mind can extend from the "infinitely Small" to "beyond the All."

The reader is taken through a mental exercise to imagine holding an object in his hand. The object of focus increases in size over eleven stages or steps. The steps include a single subatomic particle (photon), a single atom (hydrogen), a single complex biochemical (DNA), a single cell (neuron), a single organ (brain), a single person (female), a single planet (earth), a single galaxy (Milky Way), and a universe (our universe). The last two steps involve imagining that your mind is holding the previous nine steps in your hand, and then that your mind is part of a larger Universal Mind that is enabling you to do this in the first place.

The lesson of this exercise is that the human mind has the potential to extend beyond everything it can even imagine. Within the Small—our individual minds—is the infinite potential of the All.

The theoretically infinite potential of the human mind extends to understanding various types and levels of designing processes. Five levels of designing are described. Designing processes can be thought of as being (1) relatively independent, (2) relatively genetically determined, (3) relatively shaped by education and culture, (4) relatively emergent through interactions with other people, living as well as "in spirit" (discussed in *The Afterlife Experiments*), and (5) guided to some degree by a universal higher intelligence. All five levels can operate to various degrees and can be a manifestation of a universal, invisible G.O.D. process.

The chapter ends with a discussion about how the human mind can discover the so-called weird nature of light (sometimes referred to as quantum weirdness). It is explained how light is remarkably invisible to light—beams of light can pass through each other without disturbing each other. Light is able to do this presumably because it is both massless (has no mass) and spaceless (is infinitely small). It is this seemingly miraculously property of light that enables it to store infinite information in the vacuum of space.

Is this seemingly impossible nature of light potential evidence of a superintelligent designing process? Was this property of light designed for a special purpose? The answer that is revealed in the chapter is both simple and profound.

CHAPTER 13. EVIDENCE-BASED FAITH

There are five different ways that people form their beliefs and ulti- mately their faiths. The five ways are (1) education-based beliefs taught by their families and institutions, (2) emotion-based beliefs shaped by their emotions and wishes, (3) experience-based beliefs based upon their direct personal experiences and how they interpret these experi- ences, (4) reason-based beliefs formed through logic and inductive thought, and (5) evidence-based beliefs developed through personal experiments plus systematic research of scientists. The same five ways apply to the formation of faiths as well.

Our beliefs about the world—be they spiritual or otherwise—re- flect some combination of these five ways of believing.

How do you form your beliefs? Which ways do you use to deter- mine your faiths?

Particularly concerning religions, we are told that we should base our beliefs and faiths on what our parents and religious leaders tell us—i.e., education-based faith—regardless of whether or not the sto- ries we are told feel right (emotion-based), fit our direct personal expe- riences (experience-based), are logical (reason-based), or fit the facts as supported by scientific data (evidence-based).

However, if our faith does not fit the facts of nature, why should we maintain the faith? If the emperor has no clothes, shouldn't we say so?

This chapter proposes that if we use evidence-based faith as a primary way of forming our faiths, science curiously does not take away the idea of meaning, purpose, and higher intelligence in the universe; it actually supports the concept of a universal and intelligent G.O.D.

process in a way that turns out to be remarkably true to the spiritual essence, commonalities, and foundations of early religious education, emotion, experience, and logic. By "essence" I do not mean creation stories and the like told by native cultures or written in the Bible; I mean processes that reflect universal principles that can be documented scientifically.

The chapter emphasizes how, even when the evidence indicates that something appears to happen 100 percent of the time, we must nevertheless still take it on faith and trust that it will continue to happen that way again in the future. Evidence-based faith teaches us that nothing is completely certain, not even when something historically has occurred 100 percent of the time. Evidence-based faith reminds us to remain open to changing our minds in the future, to function cautiously, with humility.

The chapter ends with examples of how we can apply this to our daily lives, including applying evidence-based faith to discoveries of what the late Susy Smith said was "too coincidental to be accidental." The author shares an example from his life—an apparent car accident that totally demolished his car. Only much later did he realize the coincidental nature of how and why he could have survived the crash virtually unscathed.

CHAPTER 14. IMPLICATIONS OF INTELLIGENT EVOLUTION FOR SOCIETY

If we take an evidence-based-faith approach to the existence of some sort of universal G.O.D. process in our personal lives, the life of the planet, and the evolution of the universe as a whole, what implications does this approach have for science, education, business, law, politics, and religion?

Virtually no aspect of human activity is unaffected by this transformation in how we view the cosmos and our place in it.

This visionary and controversial chapter cannot be easily summarized—you must read it to appreciate it. One illustration is presented below from the section "Considering Feedback from G.O.D. in Business Decisions."

What do you think might happen if business were to reexamine itself, and it considered receiving guidance and direction from the universe that it ultimately serves?

If we are a product of an infinite intelligence—and I underscore *if*—the evidence indicates that we have been given substantial freedom to believe what we want, regardless of whether or not our beliefs match reality. We even have the freedom to believe or not in the existence of an infinite intelligence that potentially provided this gift of freedom in the first place! Freedom to believe, for better or worse, appears to be a purposeful part of intelligent evolution.

We have a choice not to believe in a G.O.D. process. And we can run our businesses without thinking about G.O.D. in the context of what we do. In fact, this is the way most businesses function.

However, we also have the choice to believe in a G.O.D. process. We can, if we so choose, take G.O.D. into account in the design and execution of what we do in our business lives. We can even consider consulting with people specially trained who regularly ask the universe for feedback about business choices and values, and see what information they receive—what I term "evidence-based oracles."

Should we "ask God first" when we make business decisions? Should we place this information on the boardroom table, and see how it fits with what we want to do?

Is G.O.D. the ultimate CEO? Are our CEOs ultimately under the invisible guidance? What if our CEOs were to ask for more overt guidance from G.O.D.? Would our world become a healthier, joyful, more peaceful and fulfilling home?

Science is no longer taking G.O.D. away; it is discovering G.O.D. in every place it looks. Science is not an enemy of a G.O.D. process; it is his/her/its/their ultimate servant.

Science is the universal tool by which we can know the G.O.D. process and reveal the great potential that exists within us and around us.

Science is not only enabling us to see the power of our individual minds; it is enabling us to discover the universal meta-mind that provides the spark for all of us.

CHAPTER 15. THE ORGANIZING MIND

In this chapter we return to research and address the question, can the human mind interact with subatomic particles that are isolated by electrical and magnetic shielding? Experiments are reported using an electronic device that measures the behavior of electrons in an electromagnetically shielded box. The findings indicate that when a person is in a mental state that is meditative and/or absorbing, the electromagnetically isolated electrons will begin moving away from presumed randomness. In the process, the behavior of the electrons becomes more organized.

Moreover, if groups of people are engaged in synchronized meditation and other tasks that are absorbing, the behavior of the electrons will become more organized.

Even though the people were not trying to influence the electrons, the electrons responded to the mental state of the people. The experiment involved approximately twenty-five hundred practitioners of a kind of Chinese Qigong who, save for our Chinese collaborators and their colleagues, were completely unaware that their meeting was being monitored.

The findings suggest two conclusions. The first is that the human mind, when it is in certain states, can have an organizing effect even at the subatomic level. The second is that this effect is not mediated by electromagnetic fields per se, but may involve a more direct quantum connection if not actual consciousness connection attributed to the hypothesis of universal consciousness.

CHAPTER 16. WISDOM IN THE STARS

Walter Cannon, M.D., the distinguished professor of physiology at Harvard University, wrote a book titled *The Wisdom of the Body* that addressed the question "How does the physical body become organized and maintain a state of balance and health?"

Chapter 16 takes this question further by asking whether there is evidence for *the wisdom of the system* or *the wisdom of the universe* as a whole.

The author describes a thought experiment conceived at Harvard involving the invention of the "superstomach" that illustrates how processes will go out of control when essential, guiding feedback systems are disrupted or disconnected.

If a universal G.O.D. process exists, and it has infinite intelligence and wisdom, then we should be able to discover this wisdom to the extent that we are able and willing to apply the knowledge for the purpose of healing, evolution, and transformation.

CHAPTER 17. THE GENIUS WITHIN EVERYONE

Two thousand years ago, the stoic philosopher Epictetus wrote: "When you close your doors and make darkness within, remember never to say you are alone, for you are not alone; God is within, and your genius is within."

If G.O.D. is some sort of universal Guiding-Organizing-Designing process, this implies that G.O.D. is some sort of omnipresent Genius-Organizing-Developer of the universe and everything within it.

Can science go beyond concluding that "some sort of" intelligent G.O.D. process exists, and discover more precisely what is the nature of this process?

Can science reveal details about G.O.D.?

Chapter 17 presents contemporary scientific evidence and associated theory that lead to the conclusion that consciousness is not a product of the brain, or even a creation of matter per se, but actually precedes and shapes the creation of energy and matter.

Dr. Dean Radin's book *The Conscious Universe* and Dr. William Tiller's book *Conscious Acts of Creation* provide substantial evidence supporting the thesis that consciousness is a universal process that is omnipresent to various degrees at every level of nature.

This visionary and controversial conclusion, coming from current physics and parapsychology, provides an empirical foundation for the existence of an intelligent, and evolving, Guiding-Organizing-Designing process at every level of nature.

The evidence is not going away. It is growing. The challenge is for us to come to understand and accept it, and then live our lives accordingly.

CHAPTER 18. SUMMING UP THE G.O.D. EXPERIMENTS—THE EMERGING CASE FOR INTELLIGENT EVOLUTION

This chapter reviews the logic and evidence presented in the book that leads to the conclusion that some sort of superintelligent G.O.D. process exists in the universe. The logic and evidence flow as follows.

Fact 1: Order is the rule. There is compelling evidence that order is the rule, not the exception, in nature and the universe, and that metapatterns replicate at every level in nature, from subatomic particles to clusters of galaxies, and everything in between.

Fact 2: Extraordinary order in everyday life. There are long-standing claims, buttressed by contemporary evidence in parapsychology (an example of such an experiment is presented in Part One), that indicate that levels of coincidences and synchronicities can be observed in daily life. These patterns are so extraordinary that they imply the existence of exceptional orchestration and coordination in life.

Fact 3: Extraordinary order is virtually impossible by chance alone. When conventional statistics are used to calculate the probabilities of such orders—including metapatterns—occurring by chance per se, the probabilities are extremely tiny if not infinitesimally small.

Fact 4: Experiments do not support chance as a valid explanation. When simple experiments are conducted to determine whether orders occur by chance alone, the findings come up negative. An experiment using colored sand is described that indicates that sand paintings are never created by chance alone. In the absence of a Guiding-Organizing-Designing process, sand always mixes. Moreover, when computer modeling experiments are conducted that set up the required conditions for randomness to occur—the key condition is independence of events—the computer always generates a normal curve.

Fact 5: Nature does not fit the required conditions for chance. When the required conditions for chance to occur—complete independence of events—are viewed in the context of contemporary physics, we discover that nothing that exists meets the criteria of independence. All things are interconnected and intertwined by physical fields—including gravitational and electromagnetic fields. Randomness cannot occur in the absence of complete independence.

Some sort of an intelligent, trial-and-error, Guiding-Organizing-Designing process is required to explain 1–5 above (Part Five).

Fact 6: The universe sometimes answers questions. Exploratory experiments are presented indicating that under certain conditions, it is possible to ask the universe questions and receive answers that can be verified. The findings provide proof in principle that some sort of a G.O.D. process can be accessed by the human mind and brought into the laboratory in future research (Part Three).

Fact 7: Ockham's Razor and parsimonious explanation. Although certain phenomena in nature—such as the emergence of clouds or the growing of oak trees—appear to self-organize, the key term here is "appear." The history of science reminds us that just because the earth appears to be flat, or that the sun appears to revolve around the earth, does not necessarily mean that these are the correct perceptions. Ockham's Razor is used in science to select the simplest explanation that not only accounts for the largest amount of the data, but also makes novel predictions that can be confirmed or disconfirmed in future research. When Ockham's Razor is applied to the totality of the evidence, it picks some sort of an intelligent, trial-and-error, G.O.D. process explanation.

The conventional explanation offered by skeptics, is that the existence of natural laws, plus randomness, can account for both creativity and evolution of orders in the universe. However, these scientists do not address the question about where the laws and their essential organization and compatibility come from (chance is not a viable explanation), nor how random sampling—which requires independence—could exist to foster creative orders in an interconnected universe.

Chapter 18 ends with a question: can science go beyond the "some sort of" G.O.D. explanation to discover details of the nature of G.O.D.? The answer is yes, mathematically and psychologically.

EPILOGUE. INFINITE LOVE: THE ULTIMATE GIFT FROM G.O.D.?

No discussion of an evidence-based-faith approach to G.O.D. can be complete without introducing the concept of universal infinite love as an expression of an infinite G.O.D. process. Religious writers throughout recorded history have typically described the universal God as a "God of love."

This chapter reveals how humans, more than any species, have the capacity to love virtually everything. We can love people, animals, plants, places, art, music, buildings, machines, sports, stories, and ideas themselves. We are born into the world as little spiritual bio–love machines. Our potential to love is as vast as our potential to think.

However, just as our capacity for thinking and creativity requires careful education and training, our capacity for love and compassion requires careful nurturing and modeling. Humans' capacity to love is as much a miracle as is our capacity to understand.

Just as we too often take our mind's potential for granted, we take our heart's potential for granted as well. The question is, can we learn to love wisely as we learn to think wisely?

The chapter begins with the question, is G.O.D. "too cruel to be imagined"? Or are we given choice—in other words, given the chance—to choose wisdom and love over ignorance and hatred? Who are you, ultimately—your education and history, or your potential to grow and evolve?

Who do you wish to be?

If the "All is in the Small," then our universal potential for infinite love may be the universe's ultimate gift.

The G.O.D. process says, "Ahh, finally you have the idea."

APPENDIX B

Frequently Asked Questions and Some Critical Answers

What led you to the conclusion that some sort of a G.O.D. process must exist in the universe?

I was led there first by theory, second by experiments, and third by personal experiences. I was not led there by religion or politics.

I was originally led to the G.O.D. process conclusion in the early 1980s when I read textbooks on systems theory and integrated this information with contemporary physics.

Systems (and networks) are highly interconnected, and everything is ultimately interconnected by physical fields. Being well trained in statistics, I realized that if I accepted the reality required by systems theory and physical fields—and I underscore "if"—then chance and randomness as defined statistically (the orders that occur with random sampling) could not actually operate in the universe, since the statistics of chance and random sampling require complete independence of events.

Appreciating the theoretical prediction that chance could not operate in an interconnected universe, I then began to test whether this prediction was true or not using simple experiments like the sand painting experiment, followed by computer modeling experiments.

The experiments were definitive—for example, sand paintings never occurred by chance per se; instead, sand always mixed. And computer programs designed to allow chance to occur always produced normal distributions.

My personal experiences, approached as exploratory experiments, came later. It is possible that my personal exploratory experiments turned out to be so positive because in light of the compelling theoretical and experimental discoveries, I was open to the possibility of a G.O.D. process.

Was the G.O.D. process concept based upon your religious upbringing?

Not at all.

I was raised in a Reform Jewish home that practiced what I call "devout agnosticism" and what my writing partner calls "orthodox agnosticism." Whether the question raised was about gravity or God, the approach was the same.

My parents taught me, and I paraphrase, "I don't know. Could be yes, could be no. Show me the data. I'm open." Their philosophy was not "It's impossible to know about gravity or God"; rather it was "I don't know, but I'm open to learning, one way or the other." I was the annoying child in school who was always asking questions like "Why is the sky blue" and "Why are photons invisible?"

I did go to Catholic church with girlfriends, I played basketball for a while as a member of a Methodist church team, and I learned a bit about how Christians viewed God.

Meanwhile, my Jewish upbringing spoke of a white man with a beard and cane who purportedly not only killed Egyptians but indirectly allowed his son to be killed as well.

The truth is, as an adolescent and adult I viewed such stories as myths and fables, and I rejected them.

Are you a religious person?

In the sense of holding a specific set of beliefs and following a particular practice, the answer is no.

Are you a spiritual person?

Yes, but I have come to my spirituality through science.

Are you a creationist?

Since I do not hold a specific set of religious beliefs, I cannot be a creationist—and this book is not a creationist book.

Are you a Kabbalist or a numerologist?

Same answer. I do not hold beliefs that would allow me to be a Kabbalist or a numerologist. Just because some of the data reported in this book are consistent with certain predictions of specific religious or metaphysical groups—including creationists, Kabbalists, and numerologists—that does not imply that I adopt the overall philosophies and practices of these groups.

Are you a Darwinian evolutionist or a Dembskian "intelligent designist"?

In the scientific sense I subscribe to the concepts of both evolution and intelligent design, but only when these words are defined in terms of their original, generic meanings. I am neither a Darwinian nor an Dembskian, especially in terms of the politics of science or religion.

The only "ist" that I firmly adopt is "scientist." My purpose is to be a "truthist"—concerned with the truth, whatever it is.

Were you an atheist when you began this work?

Somewhat. I was originally an agnostic, leaning toward the atheist side.

What books, early on, opened your mind to the possibility of a G.O.D. process?

Early on, the books that opened my eyes came primarily from physiology and systems theory.

Dr. Walter Cannon's *The Wisdom of the Body* opened my eyes to the existence of remarkable order in the body (and in his last chapter, of order in social systems). Cannon introduced me to the idea that feedback was a universal process in all systems at all levels.

Dr. James G. Miller's *Living Systems* opened my mind to the existence of complex systems sharing common organizations that "shred out" (as he termed it) from cells and organs to organisms and institutions. Miller showed me how orders seemed to replicate, from the micro to the macro.

Dr. Heinz Pagels's book *The Cosmic Code* opened my mind to the existence of codes (orders) that operate even at the quantum level. He also reminded me that fields are the rule, not the exception, everywhere, including the "vacuum" of space.

I read hundreds of books during that period. But these three provided the inspiration that ultimately led to my writing this book.

Is research on the G.O.D. process encouraged by your discipline or colleagues?

For the most part, no.

Psychologists, as a group, tend to focus on individuals or groups of individuals. This is their mission.

They are not typically well trained in mathematics, physics, systems science, complexity theory, parapsychology, ecology, or astrophysics. And as a group they have little interest in these topics.

Hence, the discipline of psychology does not stimulate its constituents to ponder big-picture questions or think integratively.

In addition, academic psychologists tend to be secular, not strongly religious or spiritual, and a significant subset of them are confirmed atheists. Also, evolutionary psychology is growing.

However, the psychology of religion is growing, and I created a course at the University of Arizona titled Psychology of Religion and Spirituality.

Have your conceptions been inspired by certain individuals?

Yes. Einstein has been a huge hero and role model. His interests spanned the scientific and the spiritual. Cannon, Miller, and Pagels have each inspired me in their own ways.

Contemporary senior scientists like Professors Rustum Roy and William Tiller, discussed in Chapters 16–18, as well as Drs. Radin and

Nelson as examples (see Chapter 7), are visionary scientists who see the need to integrate science, spirituality, and healing.

What was the critical theoretical insight that led you to conclude that chance could not be the primary explanation for the origin and evolution of order in the universe?

The key was the concept of the *field* in physics. Beginning with Sir Isaac Newton and his concept of the gravitational field, physics became open to inferring the existence of invisible fields that could travel in all directions and extend into space, potentially infinitely. When Sir James Clerk Maxwell developed the mathematical equations for electromagnetic fields, it became clear that massless and spaceless particles, termed photons, existed and traveled in the "vacuum" of space and interconnected everything that existed in the universe.

The truth is that fields are ultimately completely mysterious. What sense does it make to posit something that is nonmaterial (has no mass and takes up no space), yet travels in space and provides information—form—that regulates everything that is material?

Physicists posit the existence of fields because the math (logic) dictates their existence, and experiments provide data that require their existence.

Einstein said, "The field is the only reality." Einstein was driven to discover a unified field theory. He never reached his goal.

The existence of fields eliminates independence in the universe. No independence, no randomness as defined by statistics. We can't use the chance explanation, first and foremost, because the universe does not fit the conditions required for chance to occur.

What was the critical experimental insight that led you to conclude that chance could not be the primary explanation for the origin and evolution of order in the universe?

I had many experimental insights that were critical. However, my favorite involved sand paintings and sand mixing. Everyone can understand the conditions for creating sand paintings. Everyone can conduct the sand painting container experiment. Everyone can discover that sand always mixes in the absence of an ordering process. Sand paintings can be beautiful. And sand paintings are fundamentally spiritual and are used for healing by native cultures.

The sand painting experiment was not the very first experiment that led me to the deep experimental insight. The experimental insight first occurred to me in a more mundane setting—as I was folding my clothing in a lovely Maine Laundromat that looked out over the ocean.

A woman across the table from me was creating a folded clothing sculpture—and mine by comparison looked like a mess. However, despite the prediction of simple statistics that by chance alone, one's clothing could fold, if you conduct the experiment (throw your clean clothes in the air), does it ever come down neatly folded? Of course not.

If the G.O.D. process is truly universal, it must play a role not only in guiding the behavior of electrons, DNA, brains, sand paintings, people's lives, the earth's climate, the paths of the planets, the spiraling of galaxies, and the formation of superclusters of galaxies; it must also play a role in guiding even the most mundane of events, such as the folding of clothing.

The price a scientist pays for positing the existence of a universal process is that she or he has the responsibility to determine whether the claim for universality can be met or not.

What was the critical personal experience you had that led you to conclude that chance could not be the primary explanation for the origin and evolution of order in the universe?

The most powerful personal experience was probably the set of events I describe in Appendix C: "Extraordinary Synchronicity in New York City." If you could put yourself in my shoes, and ponder what I must have felt as I witnessed one improbable event after another occurring, you could imagine my sense of both wonder and fear. I could not deny seeing the patterns in the evidence, yet the totality of the evidence seemed completely unbelievable to me as well as to most people who hear the story.

Only the existence of some sort of an intelligent G.O.D. process could explain this pattern of data.

How do you respond to skeptics who claim that if you wait long enough, sand paintings will ultimately paint themselves?

My response is simple. I ask them the following question: "Why do you think that if you wait long enough, it will happen by chance?"

Is their prediction based upon data? Clearly no, because the evidence from the sand container box shows that every single time, the sand moves toward uniformity as it mixes. It never does otherwise. Never.

Is their prediction based upon theory? Clearly no, because chance statistics requires that each event be independent of every other event, and this condition does not exist in a sand container box.

So if their prediction is inconsistent with the evidence, and is not based upon logic consistent with scientific theory, then what is it based upon?

Possibilities include (1) misunderstanding statistics, (2) distrust of physics, (3) wishful thinking, (4) magical thinking, (5) need to believe that a G.O.D. process does not exist, (6) fear of invisible guidance and power implied by a G.O.D. process, (7) lack of humility and inability to accept being wrong.

It is remotely possible that they have a valid reason—however, given everything that we currently understand in math, physics, psychology, and parapsychology, I just don't know at this point what this could be.

How is your concept of evidence-based faith related to William James's concept of radical empiricism?

William James was ultimately an empiricist. Even though James loved ideas (he was a philosopher as well as a scientist), he was more in favor of evidence than ideas. I suspect that James would approve of the philosophy of evidence-based faith.

In your opinion, what is the most plausible and responsible skeptical criticism of the position you have taken in this book?

A responsible skeptic will point out that sometimes Ockham's Razor is wrong.

Ockham's Razor is not an ironclad rule; it is a helpful guide. The phrase "The simplest and most parsimonious explanation is usually the correct one" should emphasize "usually."

It is possible that at least two explanations will be required to explain the totality of order in the universe, and neither will require the existence of a universal Guiding-Organizing-Designing process.

Yes, humans engage in intelligent, trial-and-error design. So do lower animals to various degrees. Intelligence exists and is growing— artificial intelligence can now be packaged in man-made machines.

And yes, sand does not paint itself; and watches do not design themselves or self-assemble.

However, as I discuss in Chapter 12, clouds form, rains fall, oak trees grow, and humans are born. My proposal that this apparent self-organization is just "apparent" (i.e., that it requires invisible G.O.D. fields or what Sheldrake calls "morphologic fields") could be wrong, even though physics tells us that self-organization depends upon the precise modulation by patterns of external fields. Maybe instances of pure self-organization actually exist.

As a general rule, skeptics usually like Ockham's Razor. However, we must again emphasize the term "usually." In this case, it is appropriate to be open to the possibility that although the totality of the evidence best fits the parsimonious G.O.D. process explanation, this does not necessarily mean that it is the correct explanation.

Future experiments can be conducted to provide a more definitive answer.

In your opinion, what is the motivation of people who are superskeptical about the existence of some sort of a G.O.D. process in the universe?

Some people are motivated to be superskeptical because they are professional skeptics. They edit skeptical magazines, serve as officers in skeptical societies, and advocate skeptical positions in public forums.

When I gave an invited address on *The Afterlife Experiments* to the twenty-fifth-anniversary meeting of the founding of the Committee to Investigate Claims of the Paranormal (CSICOPS), I learned that approximately 95 percent of the members of the organization were devoted atheists.

If you are a motivated and devoted atheist, whether or not you are a leader of the skeptical movement, you will find the evidence for a G.O.D. process challenging if not upsetting.

However, if after reading the evidence, atheists cannot come up with a good reason to reject the evidence or its conclusion, then there are other factors besides logic and evidence playing a role in their skepticism.

I suspect superskeptics suffer from a form of what I suffer from. I call it PESD (after PTSD, posttraumatic stress disorder). "PESD" stands for posteducation stress disorder.

You will recall in Chapter 6 when I first asked the universe a question. You will remember that I hated the word "God." I asked for a new name for God because I suffered from PESD when came to the idea of a G.O.D. process, and I had already come to the conclusion that the emerging logic and evidence no longer justified my very agnostic position. It is one thing to question evidence; it is another to dismiss evidence and the reality it reveals.

Are you still an agnostic about God?

Currently I would define myself as a recovered agnostic concerning the G.O.D. question. Moreover, an unintended consequence of writing a book like this is that it becomes much harder—if not impossible—to deny the evidence and the conclusions you have reached. It is said that the proof is in the pudding. I have tasted heaping plates of G.O.D. pudding; and the taste is metaphorically divine.

One of my colleagues, sophisticated in statistics and physics, severely criticized my failure to distinguish between random sampling—a statistical technique for modeling chance processes—and randomness—an observation of apparent unpredictability that is laden with implicit interpretations (such as the process is inherently unpredictable) in an earlier draft of the book. He was correct. Constructive feedback is invaluable.

Also, for the record, when I report observations related to the

Kabbalah, it is important to know that I am not a Kabbalist, and I do not subscribe to its historical claims. Though I resonate with some of its history and ideals, I do so from the perspective of a scientist, approaching his personal life as a natural laboratory for discovery and learning.

Do you believe in a personal God?

I have come to such a belief, and my belief is evidence-based. Systems science provides the logic. Experiments in physics, mathematics, computer science, and parapsychology provide supportive evidence. Direct personal experience (received in the context of personal exploratory experiments) provides confirmatory evidence.

Why do you believe that an intelligent G.O.D. process is likely to be a creative, trial-and-error, Experimenting G.O.D.?

If we adopt systems science, and then examine the evidence from nature, particularly evolution, we witness a complex process where not only is novelty the rule (and not the exception) but also extinction is the rule (and not the exception).

If (1) nature is an expression of a G.O.D. process, and (2) some sort of natural selection process is occurring, and (3) the natural selection process is intelligent (intelligent evolution, the conclusion drawn in this book), then (4) the kind of intelligent designing process expressed is one that fits an intelligent, trial-and-error designing process.

The idea of a creative, trial-and-error Experimenting G.O.D. is an evidence-based hypothesis, not a culturally (religiously) based one.

What kind of evidence would convince you that the G.O.D. process theory was wrong?

This is a difficult and very important question. It is difficult because I don't know of a plausible alternative explanation at the present time. It is very important because one ideal of science is to be able, in principle, to disconfirm a given hypothesis.

In certain areas of science this is sometimes extremely difficult to do. Astrophysics is one such discipline where certain hypotheses and explanations cannot be experimentally disconfirmed.

Obviously, if future experiments can demonstrate that sand can actually paint itself, or that watches can assemble themselves, for example, this would be potentially convincing. Or if evidence of complete independence is discovered to exist in future research in physics (termed a completely closed system—none have been discovered to this date), this would be potentially convincing.

If some fundamental flaw in the reasoning was discovered, this would be potentially convincing.

This is why I qualify "beyond any doubt" with "virtually."

Keeping a little bit of doubt is always healthy.

Are you a G.O.D. advocate?

No. I explain to people that I am not trying to prove the existence of a G.O.D. process, nor am I trying to disprove the plausibility of chance as an explanation for order in the universe.

What I am doing is attempting to give the G.O.D. process the opportunity (or "chance") to prove itself. Experimentally we attempt to create optimal conditions to allow the G.O.D. process, if it is real, to reveal itself. Thus far, it has.

I have also conducted experiments that have attempted to give the chance explanation every possible chance to prove itself. The fact that it has repeatedly failed to do so is telling.

Are you a truth advocate?

Emphatically and unequivocally yes. Harvard's motto is *Veritas* and Yale's motto is *Lux et Veritas* ("Light and truth").

Why do you come to the conclusion that the G.O.D. process may be evolving?

The preponderance of evidence indicates that the universe has been evolving for approximately 12 billion years or more. If the G.O.D. process is intimately involved with the origin and evolution of order in the universe, it is reasonable to posit that the G.O.D. process itself could be evolving as well.

How do you respond to people who claim that they don't understand the logical leaps you make?

My responsibility, as an educator and writer, is to try to make the logic as understandable as possible. Part of the reason I work with a gifted professional writer is to increase the probability that my logical steps are understandable.

Hence, when someone does not understand the logic, I attempt to find new words and examples to help him. However, sometimes even simple logic is not understandable, that is, if the person does not want to understand.

A dear friend once confessed to me, "If I dislike something, my feelings stop me from understanding it."

It is hard to try to understand things when they make you angry, frightened, saddened, or disgusted. For some people, the idea of a G.O.D. makes them angry, frightened, saddened, or disgusted.

This helps explain why an interested child can learn this information so quickly while some hardened adults find it difficult if not impossible to understand.

How do you respond to people who claim that the logical leaps you have made are unjustified, if not wrong?

My response is simple. It is, "Please explain what the errors are that you see."

I believe in the universality of feedback and the Guiding-Organizing-Designing information that honest feedback provides. To guide includes the process of correcting.

No one that I know is perfect, and that includes me. Hence, I always ask people for their explanations.

Some of my colleagues provided me with some critical and helpful feedback that I included in the book—such as responding explicitly to conventional scientists who believe that physical and chemical laws are sufficient to account for the origin and evolution of complex biochemical and biological systems, even though they avoid the question of explaining the order and evolution of the laws and their organization in the first place. I also received some critical feedback which I thought was unnecessary to include in the book—such as "maybe some sort of clockwork universe model could be formulated in the future that did not require intelligence" (which I would consider, using his own words, to be "idle speculation" at best), or that maybe "interested spirits who were precognitive manipulated the presumably random shuffling of the envelopes by the secret person in California to create the apparent extraordinary synchronization and coordination observed in the Christopher Robinson experiment" (which is slightly less speculative than the clockwork universe model, but extremely improbable given the available evidence regarding the "power" of deceased individuals to precisely control our behavior).

A person's explanation may be correct or incorrect in a given case. Part of the fun of science is in discovering which it is.

What led you to hypothesize that a G.O.D. process involves Universal Consciousness and Intelligence?

The idea of Universal Consciousness is ancient, and I was familiar with various philosophical and religious arguments pro and con to the concept.

However, it was a former graduate student at the University of Arizona—Lonnie Nelson, Ph.D.—who pushed me to go beyond the "some-sort of" G.O.D. process explanation to address the "what kind of" G.O.D. process question.

A three-hour conversation with Lonnie led us to address the question of the universality of consciousness as implied by the idea of a universal, intelligent G.O.D. process.

The synthesis outlined in Chapter 17, written after the first polished draft of this book was completed, was formulated after that fateful conversation.

Are there any young people who inspire you to take a scientific approach to the G.O.D. question?

Thankfully, yes. Lonnie Nelson, Ph.D., is one. Others are Sheryl Attig, Julie Beischel, Ph.D., Shauna Shapiro, Ph.D., Shamini Jain, Dan Lewis, and Sabrina Lewis. Others—who seem still young to me—include Katherine Creath, Ph.D., Lewis Mehl-Madrona, M.D., Ph.D., and Katherine Burleson, M.D. The combination of their inspiration and critical examination provides me with energy and guidance.

What do you see are the core components of the coming paradigm change implied by this book?

One big change is our idea of randomness, chance, and disorder. If order is the rule in the universe, and what we term disorder is a particular kind of order (e.g., an order too complex for us to describe at the present time), then many of our perceptions in science and life will need to change accordingly.

A second big change is our idea of interconnectedness. If everything we do, including what we think, takes place in a sea of interdependence and intermixing, it follows that our sense of individuality and responsibility will need to change accordingly.

A third big change is our idea of intelligence. If intelligence is the rule in the universe, then even lowly photons and electrons may be found to have a proto-intelligence, and our science will need to discover these properties.

A fourth big change is our idea of consciousness. If consciousness is the rule in the universe, then our individual minds are expressions of a vast conscious potential, and our answers to questions about who we are and why are we here will take on deeper meaning. Also, how we treat animals, plants, and the environment will take on a new perspective.

A fifth big change is our idea of physical laws and their purpose. First, as Rupert Sheldrake writes, if everything that exists is in a state of evolution, including evolution itself, then what we call physical laws may be in evolution too (he terms them "habits" of nature). Second, since the evidence indicates that there are universal orders/laws, the question arises, how do these laws relate to the question of intelligent design? The answer is, what we call laws reflect intelligently designed principles and processes. In other words, an economical/parsimonious universal designer will, like a skilled computer programmer, create commands that can serve specific design functions.

From this perspective, just as commands are created by computer programmers, universal commands are created by the Universal Programmer.

If the universe is a conscious universe, how do we as individuals fit into the grand scheme of things?

As far as we know, only human beings can hold an image of photons in one hand and an image of the whole universe in the other. Our minds, in principle, are bigger in scope than the entire universe. If the universe is a universal, intelligent G.O.D. process, then we are little intelligent g.o.d. processes, serving the big G.O.D.

Jesus' words, which are not unique to him (and he claimed they came from the "Father"), that we are "all children of God" take on a deep systems and physical fields meaning in light of this book.

Why do you believe that many people resist the idea of a God?

The invisible can be scary. The invisible can be unpredictable. The power of a single tornado frightens us; the power of the sun is more daunting. How much power does the G.O.D. process have? Can you really imagine what "infinite power" means?

Are we powerless in the face of G.O.D.? Is G.O.D. really a caring G.O.D. when it not only allows plagues and hurricanes, but allows humans to abuse themselves and everything else around them?

And then there is all that intelligence. If G.O.D. can coordinate the entire universe, as well as relationships within our lives to various degrees, we feel pretty small intellectually in comparison.

In terms of energy, G.O.D. is huge and we are tiny. However, in terms of consciousness, our minds have a huge scope. The challenge is for us to discover this potential and use it wisely.

How can science and religion ever join forces?

The key to joining forces will likely require that they both agree to evolve as a function of what they learn.

Moses, Buddha, Jesus, and Muhammad, for example, believed (as far as we know) that the world was flat and that the sun revolved around the earth. Their ideas were shaped by the status of their knowledge.

Knowledge is growing. If the history of science teaches us anything, it should be humility.

To the extent that both scientists and religious leaders can learn humility, and agree to let evidence and knowledge advise them, then science and religion will slowly but surely come together as one.

Can the G.O.D. process be trusted?

The answer depends upon what we mean by trust. Can we trust gravity? The available evidence is yes; gravity is a stable field property of all material objects.

Can we trust the G.O.D. process to orchestrate weather storms to cleanse the planet and keep things flowing? And if we are in the way, will we be harmed? Again, the available evidence is obviously yes.

Can we trust the G.O.D. process to be evolving—at least in terms of its manifestation—and therefore that we must remain open-minded and keep on our toes? The available evidence says yes.

Can we trust the G.O.D. process to surprise us, whether we personally like it or not? The evidence seems to be yes.

Can we trust the G.O.D. process to keep secrets from us? If the G.O.D. process is intelligent, and is a good "parent," then I would hope so.

Will the G.O.D. process ever lie to us? As a scientist, I don't know. Personally, I hope not.

If G.O.D. has infinite intelligence and the infinite potential for love, why is there so much pain and evil in the world?

There are a number of logical possibilities.

One is that the G.O.D. process is evolving; the G.O.D. process today may be more mature than the G.O.D. process fifty thousand years ago.

Another is that the G.O.D. process appears to have given all systems, to various degrees, relative freedom and independence to make their own decisions. If G.O.D. has relative independence—I say relative because everything is interconnected, and therefore interconstrained, to various degrees—and we are a part of this G.O.D., then we have relative independence too. Just as a knife can be used for healing or harm, it follows that freedom can be used for good or evil.

A third is that the human potential for rage and destruction is not a hardware problem but a software problem. Our capacity for rage and killing can serve to protect our loved ones if they are being harmed, or be employed to abuse our "loved" ones if we have been harmed by our parents.

The gift of freedom is a double-edged sword. Our species may well not survive if it does not grow up and mature wisely into its hardware anatomy.

However, will G.O.D. allow us to completely destroy the planet? Is Gaia a living system with consciousness and intelligence? (The G.O.D. process explanation predicts yes.) Is she more powerful than we? Is there "wisdom of the earth" like there is "wisdom of the body"?

My hope is for the earth.

Does the available evidence provide any clues to the origin of the G.O.D. process itself?

No, not that I am aware of.

Why is the Gödel proof so important to you and this work?

Because Gödel appreciates mathematically that some things are so complex, and require a level of complexity. This requires that we accept such evidence with open-minded faith. Gödel teaches us humility. Humility may be a prerequisite for the universe to decide to reveal its secrets to us.

Is science a tool of God, and is G.O.D. the ultimate experimenter?

If what we mean by experimentation is intelligent, trial-and-error learning—not just the running of controlled, confirmatory experiments—than nature appears to be experimenting. If the universe is one great experiment, then G.O.D. would be the ultimate experimenter.

It is easier to feel for G.O.D., and wish to serve the G.O.D. process, if we see it being in the process of learning and evolution as we are.

Does the available evidence speak to whether the G.O.D. is male and/or female?

From a scientific point of view, what we call male, female, genderless, and androgynous are living systems that exist on the earth. If "the All is in the Small" and "the Small is in the All," then logic takes us to the conclusion that G.O.D. is all of these, and more. No single word—she, he, it, they—conveys this vision.

Does the available evidence speak to whether there is one G.O.D.?

What the available evidence suggests is that certain processes and principles appear to be universal—and universal implies one. For example, all masses appear to show evidence of what we label gravity. All systems appear to show evidence of what we label feedback. If there are a set of universal processes and principles, this implies some sort of universal G.O.D.

Universe, simply stated, is "uni-verse" which is "one-story."

If G.O.D. is so intelligent, beautiful, and compassionate, why would it make so many stupid, ugly, and dangerous creatures?

It is said that beauty is in the eye of the beholder. "Stupid" and "dangerous" are similarly relative terms.

Rabbits are dangerous to carrots; wolves are dangerous to rabbits; humans (with bows and arrows or guns) are dangerous to wolves. Is there intelligence to this ordering process at the level of the earth? Probably.

Rabbits seem to be more intelligent than carrots. However, would it be intelligent for carrots to hop around? What if plants refused to stay put? Would this be intelligent for the earth as a whole? I don't think so.

While I was writing these words, I saw some javelinas (they look like pigs) playing in the four-tiered waterfall behind my house that I can see from my desk. To humans, these creatures look quite ugly (and somewhat scary). However, they seem to really like one another. The mothers dote over the babies, and the males enjoy the females whenever they can. It seems intelligent to me for javelinas to find one another attractive, and for us to keep our distance from them!

Does the G.O.D. process have a sense of humor?

If one of the ways we come to know the G.O.D. process is to examine the evidence from nature, one wonders, What kind of Guiding-Organizing-Designing process would make it possible to create a stub-nosed, bowlegged, snorting, and drooling creature called a bulldog?

And where does our capacity for laughter come from? I would not be surprised if G.O.D. giggled.

Can science address the question "Is there really a messiah?"

It depends upon what you mean by messiah. For example, can the messiah or savior be a single person, a group of persons, or a system of thought?

And who is being saved, the Jews, the Native Americans, the human species, or the earth as a whole?

If the G.O.D. process exists, and its caring is truly universal, then it would want to help everything, not just the Jews, Christians, or Muslims.

The instinct to save may come not simply from our internal design (DNA) but from the Guiding-Organizing-Designing process as well.

Can science ever come to know the mind of the G.O.D. process?

The evidence from the history of science, especially as presented in this book, suggests yes. As we become more knowledgeable, sophisticated, and hopefully more deserving, this information will hopefully be revealed.

Is apparent self-organization an expression of the mind of the G.O.D. process?

In theory, the answer is yes. For example, to be self-consistent, I use the friendly acronym of the S.E.L.F.—the Supreme-Eternal-Living-Field—which playfully integrates dynamical complex systems theory, infinite set theory, and physical field theory. From this perspective, phenomena that we experience as showing self-organization can be viewed as reflecting a special case of S.E.L.F.-organization. For us to discover aspects of the hypothesized infinite mind of G.O.D., we must be willing to expand our minds accordingly. A step in this direction is to envision the G.O.D.S.E.L.F. and, in the process, smile.

The intellect has little to do on the road to discovery. There comes a leap in consciousness, call it Intuition or what you will, the solution comes to you and you don't know how or why.

APPENDIX C

Extraordinary Synchronicity in New York City

Being open to receiving information from the universe provides us with continuous surprises of all shapes and sizes.

Einstein appreciated that "the intellect has little to do on the road to discovery." He realized that he would take a "leap in consciousness" and the solution would come to him, and he didn't know "how or why." Einstein's explanation was to "call it Intuition or what you will."

Einstein—as did Carl Jung—appreciated the ubiquitous orders and patterns in the universe as well as our daily lives. According to Tyler Volk in his book *Metapatterns,* the evidence convincingly indicates that the universe consists of "patterns of patterns" that replicate themselves at all levels of nature, from the subatomic world, through our planetary world, to the world of superclusters of galaxies. These metapatterns include spheres, borders, binaries, centers, layers, and cycles.

Metapatterns are by definition not random. They are universal designs, and they provide a unifying structure for the universe as a whole.

However, there are exquisite patterns to events in our personal lives, including patterns in our relationships with others, which are so complex that they typically seem to us to be random. I underscore "seem to us" because our perception that they appear to be random is an interpretation that is most likely false.

281

Friedrich Schiller put it this way: "There is no such thing as chance; and what seems to us merest accident springs from the deepest source of destiny."

I share the following story about extraordinary synchronicity in New York City because it illustrates how important it is that we keep our eyes open to the probability that there is no such thing as coincidence. We should not confuse the existence of flexibility and relative freedom with unmediated and meaningless accident.

The extraordinary synchronicity (actually an avalanche of synchronicities) described in this chapter involves a degree of coincidence that is beyond what is mathematically described as "the emerging science of spontaneous order" in Steven Strogatz's book *Sync*. To explain this avalanche of evidence we must posit the existence of invisible intelligent guidance that interweaves our personal lives.

If any single set of events in my life opened both my mind and heart to Susy Smith's phrase "It's too coincidental to be accidental," it was this experience. The experience was like being run over by a caravan of Mack trucks. Once you have had such an experience, you don't doubt it, you never forget it, and you are forever transformed.

Remember—I metaphorically come from Missouri—I am a "show me" person. I share this personal story because it really happened, even though the evidence boggles the mind.

THE NUMBER 11 AT YALE

I love numbers. I seem to have a natural talent for playing with them. I won various awards for mathematics in my early schooling, and I received a nearly perfect grade in my senior year New Math course that included logical analysis and Boolean algebra. I also received a near perfect grade in advanced statistics in a graduate course I took at the University of Wisconsin.

For the record, I do not mean to brag. To balance the scales and keep things in perspective, you might enjoy knowing that I almost failed French in high school, and I continued to perform miserably in French in both college and graduate school. Whereas numbers have mostly been my friends, words in French have mostly been my nemesis.

However, there was a period of time when numbers seemed not to be my friends. It was the mid-1980s at Yale. I had discovered a weird pattern where the number 11, in various guises, seemed to be surrounding me.

Readers, please note—I am not a numerologist and I do not subscribe to numerology's claims. Although I will literally call a spade a spade if the numeric pattern "spade" shows up reliably in a given set of numbers, this does not make me a "spade-ologist."

When I describe the replicable pattern of the number 11 below, I report these patterns as a scientist, not as a numerologist. If science ends up supporting certain predictions of numerology, we must remember to distinguish carefully between the specific predictions that have been confirmed and the ancient philosophies and superstitions that surround them.

I recall the insight beginning by my noticing my office number, 1A, in the basement of Yale's Sterling-Strathcona-Sheffield Hall. The letter A is the first letter in the alphabet—which gives us 11.

The Psychology Building was on "Hillhouse Av" (which is how the sign was printed). It has 11 letters.

I then realized that I took Route 1A to get to my house. Another 11.

I lived at 326 Colonial Road. 326 adds up to 11.

Though I don't recall my precise phone numbers, license plates, and such today, those numbers—or combinations of numbers and letters—often added up to 11.

Even Connecticut had 11 letters.

Of course, I considered various explanations to account for the apparent anomaly of all these 11's. However, I did not have to search for 11's in my vicinity, they were literally all around me.

It was clear that a statistical anomaly was happening here. The probability of all these 11's occurring by chance alone was far less than one in 11 million.

However, though highly improbable, the numbers could have happened by chance. At the time I had no idea whether this anomaly had any significance.

"SOMETIMES YOU SAY THINGS THAT ARE PROPHETIC"

A professor's life can be very challenging. One juggles many activities and responsibilities at the same time.

One day a young African-American fourth-year medical student came to see me to ask if I would be his adviser for his dissertation (I apologize to him for not remembering his name). Unlike most American universities that require dissertations only of Ph.D. students, Yale honored the older European tradition of requiring that M.D. students do a dissertation, albeit somewhat less extensive, as well.

This young man wanted to do a dissertation that integrated quantum physics, acupuncture, and ancient African philosophy. I told him that I knew a fair amount about quantum physics, a little about acupuncture, and absolutely nothing about ancient African philosophy. However, if he wanted to meet weekly with me to discuss the writing of his dissertation, I would be happy to provide what advice I could.

Approximately six weeks into our meetings, he said, "Dr. Schwartz, every now and again, especially when you talk about systems theory, order, and patterns of numbers, what you say is prophetic."

"Prophetic?" I said. "What do you mean?"

He said. "You say things that are right out of the Kabbalah."

"The Ka Ba Wah?" I asked. I had no idea what he was talking about.

"No. The Kabbalah."

"The Kabbalah, what's that?"

"Ancient secret Jewish mysticism," he said, somewhat mistaking the facts.

"Ancient secret Jewish mysticism?" I replied. I had never been taught that there was anything mystical about Judaism.

"Yes. Sometimes what you say reflects these secret teachings."

The thought popped into my head: I wonder if this relates to my strange observations concerning the number 11. Of course I did not mention this to him.

I asked, "How can I learn about these secret teachings?"

He replied, "You are fortunate. Books are beginning to appear in English that describe these traditions. I suspect if you go to New York City, you can find a book."

Needless to say, that weekend I drove to New York City. I went to Scribner's bookstore on Fifth Avenue and looked for books on the Kabbalah.

Remember, this was the early 1980s, and there were very few published books in English about the Kabbalah then.

I can vividly recall going up the stairs to the second level, along the wall on the right side, the second shelf, and finding two books on the Kabbalah.

One was called *Kabbalah for the Layman*. I said to myself, I'm a layman! and I bought the work.

What I read took me by complete surprise. I saw that just as my student said, the philosophical tenets of Kabbalah were virtually the same as those of contemporary systems science. The book also described the ancient practice of numerology (though I have since learned that numerology is not an authentic part of Kabbalah), in which letters of the Hebrew alphabet were converted to numbers and summed to obtain a one- or two-digit number.

It turned out that the number 11 was a "master number." In fact, according to some texts, the number 11 was the number for "God."

It's important to understand that my learning of the existence of the Kabbalah occurred at least one year before I ultimately decided to be brave enough to ask the universe a question (Chapter 6).

Also, it's important to remember that I am definitely not a numerologist. What I am is a statistician who examines patterns in numbers and letters and recognizes when they are statistically beyond chance.

I told my then-wife, Jeanne, about this remarkable coincidence— the number 11, the black medical student, the Kabbalah, and then numerology. I told her I wanted to study the Kabbalah.

Jeanne said, "Absolutely no." She reminded me that I was a Yale professor and a highly visible scientist, not a Jewish academic mystic.

Not understanding the teachings of the Kabbalah, she was also frankly worried about how I would cope with what I might learn.

I could not argue with her. Jeanne was a smart, loving, and tough woman. However, I felt the strong need to stay connected, somehow, to this seemingly synchronous and meaningful set of events.

A few weeks later, the seemingly impossible happened. Emotionally it felt like a miracle. It still does.

A practical yet seemingly ridiculous idea popped into my head

We were in New York City again. We were staying at the Yale Club; though I can't remember the precise room number, I do remember that it added up to 11.

It was a Saturday. We were to spend the afternoon at the Metropolitan Opera. We were having lunch on Sixty-fifth Street (6+5=11). We were in a restaurant called Shun Lee West (it has a total of 11 letters).

I was silently thinking about the eerie frequency of the number 11 in New York City, wondering if my experience was due to what used to be called the "Volkswagen Bug effect"—if you think about purchasing a Volkswagen, you start noticing more Bugs on the street simply because you are now conscious of them. Or was it something more?

And then I thought to myself, How am I going to stay connected to the potential coincidence of the master number 11, the Kabbalah, numerology, and prophecy in my life?

I didn't consciously and purposely ask the universe the question. I simply asked the question silently in my head.

Immediately a wild idea popped into my head, seemingly out of nowhere, and I heard, "Get a Cardigan Welsh corgi."

Get a Cardigan Welsh corgi? As soon as I pictured the idea, I couldn't help suppressing a huge grin.

First, I immediately realized that the name Cardigan Welsh corgi had connections to the number 11. The initials *CWC* equal 3+23+3= 29=11. The letter *C* is especially important, because Kabbalah is sometimes spelled Cabbalah. *K* is the eleventh letter, C is the third letter. But in binary, 3 is expressed as 11.

There were other connections to the number 11. They are not important here. What is important is that I realized that the name of this extremely rare breed of dog could serve as a secret reminder to me of the Kabbalah as well as the number 11.

Second, I immediately realized that getting a Cardigan Welsh corgi would be acceptable to Jeanne. At that time we had a Pembroke Welsh corgi named Thurber (after James Thurber). Jeanne was about to begin working at Yale, which meant that Thurber would have to be alone during the day. We did not have children; Thurber was like a child to Jeanne and me. If we got Thurber a younger brother or sister, then he would have someone to keep him company when we were at work.

Pembroke Welsh corgis are fairly rare. When we had our Pembroke, the queen of England had eleven Pembrokes (no joke). Pembrokes have no tails, Cardigans have tails. I had once met a Cardigan Welsh corgi on Bailey Island in Maine. Though the dog's name was Killer he was a pussycat. Cardigans often have gentle dispositions—perfect for Thurber.

One almost never bumps into Cardigan Welsh corgis on the street. I mean literally almost never.

Without telling Jeanne what was going through my mind, I brought up the subject of Thurber being alone, and that maybe we should get him a friend, a Cardigan Welsh corgi.

Jeanne thought it was a great idea. She suggested that we attempt to rescue an adult Cardigan that was looking for a home. She suggested that since Cardigans are so rare, we should see if there was a Cardigan Welsh corgi club somewhere in the United States and we could let it be known that we were looking to adopt an adult Cardigan who needed a home.

Jeanne saw my excitement. She didn't know, of course, that if we could somehow, someday, find an adult corgi to rescue, he or she would become my secret Kabbalah corgi. That was my sacred secret.

What happened next was beyond belief. As I write these words, I still find it virtually impossible to believe that this actually happened.

We left the restaurant and walked uptown. We purchased two yogurt cones, and then began crossing Seventh Avenue. Around Seventy-fourth Street (yes, 7+4=11), Jeanne asked me: "Gary, is that a Cardigan Welsh corgi?"

I couldn't believe what I was seeing. In the middle island separating the uptown and downtown streets was the funniest-looking dog I had ever seen. He was a blue merle and a "fluffy"—which is a flaw for the breed. One of his eyes was brown, the other eye was blue. The brown eye had only half a pupil. He had huge ears. He had the typical short legs and long body of a corgi, with a large fluffy tail. He looked positively goofy.

How could this be happening? Out of the blue, an idea pops into my head that I should get a Cardigan Welsh corgi as a secret Kabbalah corgi, and a few minutes later I bump into a Cardigan corgi on the streets of New York City?

I fell to my knees and gave the dog my yogurt. The woman walking the dog looked a bit alarmed. However, I was wearing a jacket and tie for the opera, so I looked presentable.

I blurted out, "Please excuse me. My name is Dr. Schwartz, and you'll never guess what happened. We live in Connecticut and have a Pembroke Welsh corgi named Thurber. Just a few minutes ago we decided that we wanted to get a Cardigan Welsh corgi as a friend for Thurber. We would like to adopt an adult dog who is looking for a home, and we were just talking about how we would go about finding a Cardigan Welsh corgi club. Do you happen to know if there is such a club?"

I saw tears welling up in her eyes. What she said brought tears to my eyes. She said, "My God, my prayers have finally been answered. You may find this hard to believe. I am getting a divorce and losing my country home in Connecticut. I have an apartment not far from here. It has become necessary for me to find a home for my dog, Willie. Would you like to talk with me about possibly adopting him?"

There are no words to describe what I felt at that moment. The mixture of emotions included awe, wonder, sadness (for her), joy, pain, confusion, shock. I was frankly stunned.

I asked myself, What is the probability that this could happen by chance? How many dogs are there on the planet? How many are actually Cardigan corgis, looking for a home, who are walking at precisely the place and time that I am thinking about adopting one? We are talking about a truly infinitesimally small number.

Jeanne and I made an appointment to meet with the woman and Willie the following morning. I hardly heard the opera that afternoon. And I don't know how I slept that night.

The next morning we drove uptown to her apartment. I don't remember what street it was on. But I vividly recall the number of her apartment—5F. *F* is the sixth letter in the alphabet: 5+6=11.

It turned out the woman was an assistant opera coach at the Metropolitan Opera. She interviewed us for about an hour. She really loved Willie, and I sensed that the coincidence of us being there was almost as powerful for her as it was for me.

When the woman began to see that Jeanne and I were serious, I gently asked her, with no warning, "Do you know anything about numbers?" I just felt the need to ask. Jeanne looked at me cross-eyed. "Where did this come from?" was written on her face.

Meanwhile, the woman looked squarely in my eyes, searchingly and knowingly, and said, "Ah. Now I understand. I think you should come with me into the next room."

I followed her. What I saw I could not believe. What I was seeing was as impossible for me to believe as seeing Willie for the first time. We were in her bedroom. The long wall had floor-to-ceiling bookshelves. The entire collection of books was on astrology and numerology. Besides being an opera coach, the woman was a professional astrologer!

No joke. If you find it hard to believe what you are now reading, you can imagine how I felt. She went to the right side and from a waist-level shelf pulled out a book with the words "astrology" and "Kabbalah" in the title. I asked her if I could borrow the book. Jeanne looked at me completely confused.

The woman insisted on driving out to our home in Guilford, Connecticut. As she was getting Willie ready for the trip, I sheepishly asked her, "Do you know what the number 11 means?"

The woman looked up and said, "You know, the number 11 is a very important number."

I said, "Do you know what your apartment is?"

She said, "Of course: 5F."

I said, "5F equals 11."

Again our eyes locked. We both knew that Willie was moving to Guilford.

THE 11 CONNECTION CONTINUED, EVEN AFTER WILLIE DIED

From that day forward, Willie became my secret Kabbalah corgi. It is important to understand that Willie was not known for his intelligence. One of my former Yale Ph.D. students, Geoffrey Ahern, now an M.D., Ph.D., and professor of neurology and psychology at the University of Arizona, and a fellow dog lover, used to say that Willie operated on "two neurons."

I, of course, didn't mind his reference to "two." Geof didn't know about the 11 connection, and therefore could not know that 2 (that is, 1+1) reminded me of the master number.

Of the many dogs I have had the privilege to share my life with, Willie was unequivocally my soul mate doggie.

The connection between the Kabbalah, numbers, and Willie emerged seemingly impossibly once again, of all places, at his funeral.

It was the mid-1990s. Jeanne and I were divorced. Willie was with me in Tucson, Arizona. He had become frail; he was almost fourteen years old. I had just rushed home from a visit to Boca Raton, Florida. Debbie, a woman who stayed with Willie when I traveled, had called me explaining that Willie was behaving as if he was in severe pain. He could hardly walk.

I took Willie to an emergency animal hospital, where he was diagnosed to be in kidney failure and could not be treated. The doctor recommended that he should be let go. I held Willie in my arms as he died.

My grief was severe. It went way beyond just Willie, the dog. As Willie was dying in my arms, I realized that my living secret symbol of the Kabbalah was dying too.

The doctor offered to have someone from his staff take Willie's body to the local animal cemetery, where one of my other corgis was buried. I explained that I knew the owner of the cemetery, and would prefer to take Willie myself.

When I arrived at the cemetery, I learned that it had been recently sold to a young couple. I was introduced to the wife. I explained that I had another dog already buried in her cemetery. As we were preparing the paperwork, she asked if I wanted a footstone.

I explained that I had previously prepared my epitaph for Willie's plaque and that it was on file. She asked me to tell her the epitaph.

I said, "In Loving Memory of Willie, My Kabbalah Corgi, My Soul Mate Doggie."

She replied, surprised, "Did you say Kabbalah?"

I said, "Yes."

She said. "That's remarkable. You must meet my husband."

I could not believe what I was hearing. Willie had come to me via the Kabbalah. Could it be possible that he would be leaving me via the Kabbalah as well? Not a chance.

She went to get her husband. It turned out that not only did this young man, a Catholic in his early twenties, have an absolute passion for the Kabbalah, but he was also an expert in numerology. Moreover, with two of his high school friends who also shared a passion for Kabbalah, they had discovered the secret for calculating the solution to what is known mathematically as the "magic cube."

The details of the magic cube are not important here. What is important is that this extremely difficult numeric calculation was revealed to this man and his friends through their study of numerology in the Kabbalah.

I spent almost three hours with this young man—he came to appreciate my tears as I confessed to him why Willie was my secret Kabbalah corgi.

A few days later Willie was buried. We had a formal ceremony. It was led by a young Catholic Kabbalist who knew much more about numerology than I.

As Willie's casket was being lowered into the ground, I asked myself, How many pet cemeteries in the United States are headed by a Kabbalist who has mastered the sacred art of numerology? and, What is the probability that a Kabbalah corgi from New York City would be buried by a Kabbalah Catholic in the city of Tucson? Give me a break!

Though I had lost my soul mate doggie, I had been given a definitive reminder of a seemingly impossible gift—hard evidence that our lives are so much more than a random walk in a competitive struggle for mere survival.

My experience with Willie clearly fits the definition of "amusing" (remember Wigner in Chapter 8)—it is absolutely worth thinking about.

I share this mind-boggling experience in *The G.O.D. Experiments* not only to honor my sacred teacher and fluffy friend Willie, but to honor an intelligent G.O.D. universe that appears to be teeming with invisible Guidance, Organization, and Design. Every now and again we are privileged to be clobbered by coincidence and witness an apparent G.O.D. process in our personal lives.

The patterns are here—so too are the metapatterns.

The question is, can we listen? Will we accept? Can we understand? Shall we transform? And will we celebrate?

I hope so.

To see or not to see, that is the question.

Index

Order *(cont.)*
  ubiquitous nature of, 199–203
Organizing fields, 115–24, 119, 261. See
  also Fields; Self-organizing Systems
Organizing mind, 169–75
Origin of universe, 79–80, 95–100

Pagels, Heinz
  cosmic codes and, 141
  nonrandom quantum codes and, 122
  on probability theory, 39
  randomness and absence of rules, 182
  writings of, 46–47, 91–92, 264
Paradigm shifts, 99, 169, 174, 274–75
Parapsychology, 8–26, 191–92, 199
Pasteur, Louis, 29
Patterns, 198, 230, 281–82
*Patterns in the Void: Why Nothing Is
  Importan*t (Odenwald), 106, 119,
  228–29
Pavlov, I., 68
PDP11/GT40 computer, 134
PEAR. *See* Princeton Engineering
  Anomalies Research (PEAR)
Pearsall, Paul, 170, 173
Penn State University, 183
Perakh, Mark, 226
Personal computers. *See* Computers
PESD. *See* Posteducation stress disorder
  (PESD)
Phi, 198
Photons. *See* Light and photons
Pi, 51–52, 97, 205–6
Pledge of allegiance, 155–56
Politics, 162, 163–64
Posteducation stress disorder (PESD),
  269
Prayer, 160, 174
Precognition, 21
Presence of the Past (Sheldrake), 72
Princeton Engineering Anomalies
  Research (PEAR), 169
Princeton Theological Seminary, 228

Princeton University, 47, 92, 169, 172,
  192–93
*Privileged Planet: How Our Place in the
  Cosmos Is Designed for Discovery*
  (Gonzalez), 228
Probability theory, 39, 97
Prophetic dreams, 7–26
Proto-intelligence, 274
Proust, Marcel, 40
Psychology, 264
Psychophysiology of healing, 91
Purdue University, 227

Qigong, 172
Quantum field physics
  chance and, 96
  medicine and healing and, 91–92
  nonrandom quantum codes, 122
  Pagels and, 46–47, 91
  randomness and, 46–47
  superconductive monatomic gold and,
    233–34
  superinterdependence and, 107–8
Quantum weirdness, 172

Radiation, 79–80
Radical empiricism, 267
Radin, Dean, 172, 191, 229–30, 264–65
Radio waves, 138
Rage, 277
Random event generator (REG), 169–74
Random order, 40, 274
Random sampling, 41, 261, 269–70
Randomness. *See also* Chance
  alternative explanation for, 99–100
  apparent randomness, 101–12
  chance and, 52–53, 96–97
  conditions necessary for, 102–4, 188,
    202–3
  disorder and, 40
  DNA development and, 34–35
  evolution of order and, 205–6

About the Authors

GARY E. SCHWARTZ, Ph.D., is Professor of psychology, surgery, medicine, neurology, and psychiatry at the University of Arizona, and director of the university's Laboratory for Advances in Consciousness and Health. He received his Ph.D. from Harvard University in 1971 and was an assistant professor there for five years. He then served as a professor of psychology and psychiatry at Yale University, as well as director of the Yale Psychophysiology Center and codirector of the Yale Behavioral Medicine Clinic, before moving to Arizona in 1988. He has published more than 450 scientific papers, edited eleven academic books, and is the coauthor with Linda Russek, Ph.D., of *The Living Energy Universe* and with William L. Simon of *The Afterlife Experiments* and *The Truth About "Medium."*

WILLIAM L. SIMON is a screenwriter and the award-winning author/coauthor of more than a dozen books including *iCon*, the *New York Times* bestselling biography of Steve Jobs; *The Art of Deception,* a runaway international bestseller of fictionalized stories about computer security breaches; and *High Velocity Leadership*, the story of the Mars Pathfinder mission. As noted above, this is his third book with Gary E. Schwartz.